TestSMART

for
Math Operations & Problem Solving

Grade 3

Help for
Basic Math Skills
State Competency Tests
Achievement Tests

by

Lori Mammen

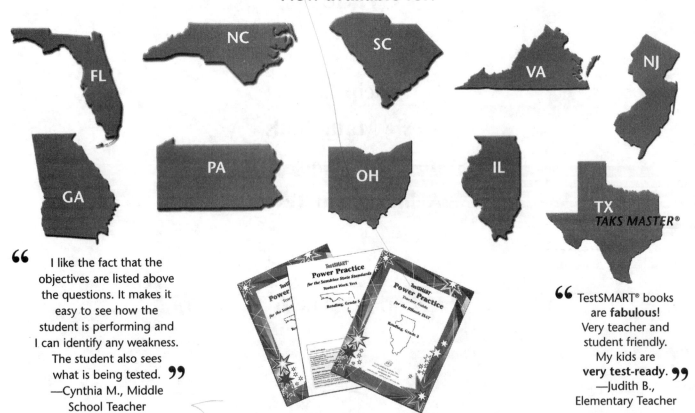

Cover: Kirstin Simpson   **Book Design:** Educational Media Services

ISBN-10: 1-57022-241-X   ISBN-13: 978-1-57022-241-2

Contents

> ## Welcome to *TestSMART®*!!
>
> It's just the tool you need
> to help students review important mathematics skills and
> prepare for standardized mathematics tests!

Introduction

During the past several years, an increasing number of American students have faced some form of state-mandated competency testing in mathematics. While several states use established achievement tests, such as the Iowa Test of Basic Skills (ITBS), to assess students' achievement in mathematics, other states' assessments focus on the skills and knowledge emphasized in their particular mathematics curriculum. Texas, for example, has administered the state-developed Texas Assessment of Academic Skills (TAAS) since 1990. The New York State Testing Program began in 1999 and tests both fourth- and eighth-grade students in mathematics.

Whatever the testing route, one point is very clear: the trend toward more and more competency testing is widespread and intense. By the spring of 1999, 48 states had adopted some type of assessment for students at various grade levels. In some states, these tests are "high-stakes" events that determine whether a student is promoted to the next grade level in school.

The emphasis on competency tests has grown directly from the national push for higher educational standards and accountability. Under increasing pressure from political leaders, business people, and the general public, policy-makers have turned to testing as a primary way to measure and improve student performance. Although experienced educators know that such test results can reveal only a small part of a much broader educational picture, state-mandated competency tests have gained a strong foothold. Teachers must find effective ways to help their students prepare for these tests—and this is where *TestSMART®* plays an important role.

What's inside this book?

Designed to help students review and practice important reading and test-taking skills, *TestSMART®* includes reproducible practice exercises in the following areas—

- pretests for each of the five major objectives addressed in the book
- practice exercises that target the specific skills tested within each objective

In addition, each *TestSMART®* book includes—

- a master skills list based on mathematics standards of several states
- complete answer keys

The content of *TestSMART®* is outlined below.

Major Objectives: This book focuses on five major mathematics objectives which represent broad areas of understanding generally common to all grade levels. These objectives focus on the students' ability to—

- use the basic operations (addition, subtraction, multiplication, and division) to solve problems
- estimate solutions to problem situations
- determine solution strategies and analyze/solve problems
- express or solve problems using mathematical representation
- evaluate the reasonableness of a solution to a problem situation

Specific Skills: A list of specific skills appears below each major objective. These skills represent appropriate grade-level expectations for a given objective.

Pretests: Five pretests are included in this book. Each pretest addresses one of the major objectives and includes test items for all the specific skills included with that objective. Teachers use the pretests to diagnose the students' areas of strength and weakness for a given objective.

Practice Exercises: Practice exercises follow each pretest. Unlike the pretests, each practice exercise addresses a specific skill. Teachers use the practice exercises to target specific areas of weakness revealed in the pretests.

Master Skills List/Correlation Chart: The mathematics skills addressed in *TestSMART®* are based on the mathematics standards and/or test specifications from several different states. No two states have identical wordings for their skills lists, but there are strong similarities from one state's list to another. The Master Skills List for Mathematics (page 7) represents a synthesis of the mathematics skills emphasized in various states. Teachers who use this book will recognize the skills that are stressed, even though the wording of a few objectives may vary slightly from that found in their own state's test specifications.

The Master Skills Correlation Chart (page 8) offers a place to identify the skills common to both *TestSMART®* and a specific state competency test. To show how such a correlation can be done, the author has included a sample correlation which shows the skills addressed in both *TestSMART®* and the skills tested on the Texas Assessment of Knowledge and Skills (TAKS).

Answer Keys: Complete answer keys appear on pages 121–124.

5

How to Use This Book

Effective Test Preparation: What is the most effective way to prepare students for any mathematics competency test? Experienced educators know that the best test preparation includes three critical components—

- a strong curriculum that includes the content and skills to be tested
- effective and varied instructional methods that allow students to learn content and skills in many different ways
- targeted practice that familiarizes students with the specific content and format of the test they will take

Obviously, a strong curriculum and effective, varied instructional methods provide the foundation for all appropriate test preparation. Contrary to what some might believe, merely "teaching the test" performs a great disservice to students. Students must acquire knowledge, practice skills, and have specific educational experiences which can never be included on tests limited by time and in scope. For this reason, books like *TestSMART®* should **never** become the heart of the curriculum or a replacement for strong instructional methods.

Targeted Practice: *TestSMART®* does, however, address the final element of effective test preparation (targeted test practice) in the following ways—

- *TestSMART®* familiarizes students with the content usually included in competency tests
- *TestSMART®* familiarizes students with the general format of such tests

When students become familiar with both the content and the format of a test, they know what to expect on the actual test. This, in turn, improves their chances for success.

Using *TestSMART®*: Used as part of the regular curriculum, *TestSMART®* allows teachers to—

- pretest skills needed for the actual test students will take
- determine students' areas of strength and/or weakness
- provide meaningful test-taking practice for students
- ease students' test anxiety
- communicate test expectations and content to parents

Master Skills List

I. Use the operation of addition to solve problems
A. Add or model addition using pictures, words, and numbers
B. Identify the number sentence that shows the commutative property of addition
C. Add simple fractions
D. Add money with and without models

II. Use the operation of subtraction to solve problems
A. Subtract or model subtraction using pictures, words, and numbers
B. Subtract simple fractions
C. Subtract money with and without models

III. Use the operation of multiplication to solve problems
A. Solve and record multiplication problems (one-digit multiplier)

IV. Use the operation of division to solve problems
A. Use models to solve simple division problems
B. Solve division problems with multi-digit dividend (one-digit divisor)
C. Determine unit cost when given total cost and number of units

V. Estimate solutions to a problem situation
A. Estimate sums and differences beyond basic facts

VI. Determine solution strategies and analyze or solve problems
A. Select and use addition or subtraction to solve problems
B. Select multiplication (one-digit multiplier) or division (one-digit divisor) and use the operation to solve problems
C. Measure to solve problems involving length, area, temperature, and time

VII. Express or solve problems using mathematical representation
A. Use appropriate mathematical notation and terms to express a solution
B. Interpret information from pictographs and bar graphs

VIII. Evaluate the reasonableness of a solution to a problem situation
A. Use problem-solving skills to solve problems and evaluate reasonableness of a solution

Master Skills Correlation Chart

Use this chart to identify the *TestSMART®* skills included on a specific state competency test. To correlate the *TestSMART®* skills to a specific state's objectives, find and mark those skills common to both. The first column shows a sample correlation based on the Texas Assessment of Knowledge and Skills (TAKS).

	Sample Correlation	
I. Use the operation of addition to solve problems		
A. add or model addition using pictures, words, and numbers	★	
B. identify the number sentence that shows the commutative property of addition		
C. add simple fractions		
D. add money with and without models		
II. Use the operation of subtraction to solve problems		
A. subtract or model subtraction using pictures, words, and numbers	★	
B. subtract simple fractions		
C. subtract money with and without models		
III. Use the operation of multiplication to solve problems		
A. solve and record multiplication problems (one-digit multiplier)	★	
IV. Use the operation of division to solve problems		
A. use models to solve simple division problems	★	
B. solve division problems with multi-digit dividend (one-digit divisor)		
C. determine unit cost when given total cost and number of units		
V. Estimate solutions to a problem situation		
A. estimate sums and differences beyond basic facts	★	
VI. Determine solution strategies and analyze or solve problems		
A. select and use addition or subtraction to solve problems	★	
B. select and use multiplication (one-digit multiplier) or division (one-digit divisor) to solve problems	★	
C. measure to solve problems involving length, area, temperature, and time	★	
VII. Express or solve problems using mathematical representation		
A. use appropriate mathematical notation and terms to express a solution	★	
B. interpret information from pictographs and bar graphs	★	
VII. Evaluate the reasonableness of a solution to a problem situation		
A. Use problem-solving skills to solve problems and evaluate reasonableness of a solution	★	

Using Addition to Solve Problems

I. Use the operation of addition to solve problems

A. Add or model addition using pictures, words, and numbers
B. Identify the number sentence that shows the commutative property of addition
C. Add simple fractions
D. Add money with and without models

Notes

Objective 1: Pretest

I.A Add or model addition using pictures, words, and numbers (1-5)

1. Darla had 7 marbles. She won 8 more during a game with her friends. Which one shows how many marbles Darla had in all?

0 **A**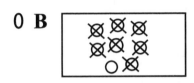

0 **B**

0 **C**

0 **D**

2. 456
    + 29
    _____

0 **A** 575

0 **B** 495

0 **C** 485

0 **D** 433

3. 627
    + 142
    _____

0 **A** 979

0 **B** 879

0 **C** 869

0 **D** 769

4. Beth has 27 seashells from the beach. Her sister has 18 seashells. How many seashells do they have in all?

0 **A** 55

0 **B** 45

0 **C** 35

0 **D** 31

5. Nancy and her mother made 12 cupcakes, 24 cookies, and 7 pies for a school fair. How many pies and cupcakes did they make?

0 **A** 19

0 **B** 31

0 **C** 36

0 **D** 43

I.B **Identify the number sentence that shows the commutative property of addition (6-9)**

6. Which names the same number as $4 + 8 + 3$?

  0 **A** $4 \times 8 + 3$

  0 **B** $4 + 8 - 3$

  0 **C** $3 + 4 + 8$

  0 **D** $8 + 3 - 4$

7. Which number would make this number sentence correct?

$16 + 7 + \boxed{} = 7 + 8 + 16$

  0 **A** 31

  0 **B** 16

  0 **C** 15

  0 **D** 8

8. Darren has 12 books in box one, 15 books in box two, and 31 books in box three. Which shows how many books Darren has?

12	15	31
One	Two	Three

  0 **A** $31 - 15 + 12$

  0 **B** $31 + 15 - 12$

  0 **C** $31 + 15 + 12$

  0 **D** $12 \times 15 + 31$

9. Which number would make this number sentence correct?

$17 + \boxed{} + 19 = 19 + 17 + 5$

  0 **A** 2

  0 **B** 5

  0 **C** 22

  0 **D** 24

I.C **Add simple fractions (10-14)**

10. Damon and Mike shared a candy bar. Each boy ate $\frac{1}{3}$ of the candy bar. How much did they eat in all?

  0 **A** $\frac{1}{6}$

  0 **B** $\frac{2}{4}$

  0 **C** $\frac{2}{3}$

  0 **D** $\frac{3}{4}$

11. A game spinner had 8 equal sections. Tori drew a bell on 1 section. Tina drew a bell on 2 more sections. How much of the spinner was marked with a bell?

0 **A** $\frac{1}{2}$

0 **B** $\frac{1}{8}$

0 **C** $\frac{3}{8}$

0 **D** $\frac{5}{8}$

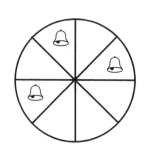

12. Mr. Davis gave $\frac{1}{5}$ of his money to his son. He gave $\frac{2}{5}$ of his money to his daughter. How much of his money did Mr. Davis give away?

0 **A** $\frac{3}{10}$

0 **B** $\frac{3}{5}$

0 **C** $\frac{1}{5}$

0 **D** $\frac{1}{10}$

13. $\frac{2}{7} + \frac{1}{7} =$

0 **A** $\frac{3}{14}$

0 **C** $\frac{3}{7}$

0 **B** $\frac{2}{14}$

0 **D** $\frac{3}{10}$

14. $\frac{1}{4} + \frac{2}{4} =$

0 **A** $\frac{2}{8}$

0 **C** $\frac{3}{8}$

0 **B** $\frac{2}{16}$

0 **D** $\frac{3}{4}$

I.D Add money with and without models (15-20)

15. Fran spent 42¢ for candy and 16¢ for a pencil at the school store. How much did Fran spend in all?

0 **A** 68¢

0 **B** 66¢

0 **C** 58¢

0 **D** 26¢

16. Danny had 45¢. He found the coins below. How much money did Danny have in all?

  0 **A** 71¢

  0 **B** 61¢

  0 **C** 26¢

  0 **D** 19¢

17. At the fair, a ride ticket costs 35¢. How much would you spend to ride two rides?

  0 **A** 50¢

  0 **B** 60¢

  0 **C** 70¢

  0 **D** 80¢

18. 37¢ + 54¢ =

  0 **A** 81¢

  0 **B** 83¢

  0 **C** 91¢

  0 **D** 93¢

19. Kathy had 67¢ in her bank. She dropped in the coins below. How much was in her bank then?

  0 **A** $1.00

  0 **B** 95¢

  0 **C** 90¢

  0 **D** 85¢

20. How much money is in the two boxes?

  0 **A** 95¢

  0 **B** 85¢

  0 **C** 83¢

  0 **D** 73¢

Practice 1.A1

I.A   Add or model addition using pictures, words, and numbers

1. David's class used 5 boxes of tissues in April and 6 boxes of tisses in May. Which one shows the total number of boxes the class used in those two months?

   0 **A**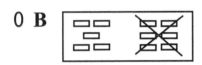

   0 **B**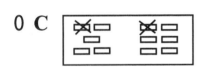

   0 **C**

   0 **D**

2.
   ```
     8 9
   + 6 9
   ```

   0 **A** 168

   0 **B** 158

   0 **C** 148

   0 **D** 129

3.
   ```
     8 7 6
   +   4 9
   ```

   0 **A** 833

   0 **B** 845

   0 **C** 925

   0 **D** 945

4. Luis had 72 stickers in his collection. His teacher gave him 19 more stickers. How many stickers did he have then?

   0 **A** 92

   0 **B** 91

   0 **C** 81

   0 **D** 61

5. Isabel planted 37 bean seeds. Her father planted 98 bean seeds. How many bean seeds did they plant in all?

   0 **A** 61

   0 **B** 124

   0 **C** 125

   0 **D** 135

Practice 1.A2

I.A Add or model addition using pictures, words, and numbers

1. Mrs. Perez had only 18 folders and 12 computer disks for her class to use. She ordered 12 more folders, so each student could have one. How many folders did she use in all?

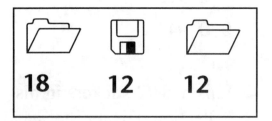

18   **12**   **12**

0 **A** 20

0 **B** 26

0 **C** 30

0 **D** 42

2.   9 7 4
      + 9 6
      ———

0 **A** 1,070

0 **B** 970

0 **C** 962

0 **D** 960

3.   5 6 1
      + 5 2 9
      ———

0 **A** 1,090

0 **B** 1,082

0 **C** 1,080

0 **D** 1,072

4. A store sold 26 pencils, 38 pens, and 19 notebooks on the first day of school. How many pencils and pens did the store sell?

0 **A** 12

0 **B** 52

0 **C** 64

0 **D** 83

5. In a city, 48 families live on Maple Street, 75 families live on Elm Street, and 63 families live on Oak Street. How many families live on the three streets in all?

0 **A** 109

0 **B** 138

0 **C** 174

0 **D** 186

Practice 1.A3

I.A Add or model addition using pictures, words, and numbers

1. Leslie mailed 9 party invitations on Monday and 6 on Tuesday. Which one shows how many she mailed in all?

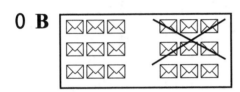

0 **A**

0 **B**

0 **C**

0 **D**

2. 1, 0 9 4
    + 9 3 7
    —————

0 **A** 1,921

0 **B** 1,931

0 **C** 1,963

0 **D** 2,031

3. 5 6 4
    + 8 7
    —————

0 **A** 477

0 **B** 541

0 **C** 641

0 **D** 651

4. The Martin family drove 345 miles on Monday and 468 miles on Tuesday. How many miles did they drive both days?

0 **A** 1,013

0 **B** 813

0 **C** 803

0 **D** 123

5. In May, 468 people ate at City Diner. In June, 566 people ate at the diner. During May and June, how many people ate there in all?

0 **A** 98

0 **B** 924

0 **C** 934

0 **D** 1,034

Practice 1.A4

1. Sheena has 42 pictures in one scrapbook and 29 pictures in another scrapbook. How many pictures does she have in all?

    0 **A** 13

    0 **B** 61

    0 **C** 67

    0 **D** 71

2. 2, 2 4 9
    + 3 3 8
    _____

    0 **A** 2,111

    0 **B** 2,577

    0 **C** 2,587

    0 **D** 2,687

3. 6 3 7
    + 7 5
    _____

    0 **A** 712

    0 **B** 702

    0 **C** 642

    0 **D** 612

4. Mr. Ramirez read 128 pages on Monday and 109 pages on Tuesday. How many pages did he read on those two days?

    0 **A** 337

    0 **B** 237

    0 **C** 227

    0 **D** 121

5. In March, 2,319 books were checked out of the city library. In April, 2,743 books were checked out of the library. How many books were checked out during those two months?

    0 **A** 4,052

    0 **B** 4,436

    0 **C** 5,062

    0 **D** 5,162

6. 1, 1 5 6
    + 8 4 4
    _____

    0 **A** 2,090

    0 **B** 2,000

    0 **C** 1,990

    0 **D** 1,900

Practice 1.B1

I.B   Identify the number sentence that shows the commutative property of addition

1. Which names the same number as 7 + 9 + 6?

   0 **A** 9 – 7 + 6

   0 **B** 6 x 7 + 9

   0 **C** 7 + 9 – 6

   0 **D** 9 + 6 + 7

2. Which number would make this number sentence correct?

   32 + 4 + ☐ = 4 + 9 + 32

   0 **A** 36

   0 **B** 19

   0 **C** 13

   0 **D** 9

3. Which names the same number as 24 + 17 + 31?

   0 **A** 17 + 31 + 24

   0 **B** 31 + 17 – 24

   0 **C** 31 – 24 + 17

   0 **D** 24 x 17 + 31

4. Which number would make this number sentence correct?

   51 + ☐ + 13 = 13 + 51 + 9

   0 **A** 9

   0 **B** 22

   0 **C** 64

   0 **D** 73

5. Which names the same number as 14 + 82 + 33?

   0 **A** 14 x 82 + 33

   0 **B** 14 + 82 – 33

   0 **C** 33 + 14 + 82

   0 **D** 82 + 33 – 14

6. Suzanne has three pails of shells. She has 26 shells in pail one, 34 shells in pail two, and 18 shells in pail three. Which one shows how many shells she has?

| 26 | 34 | 18 |
| One | Two | Three |

   0 **A** 26 x 34 + 18

   0 **B** 26 + 34 – 18

   0 **C** 18 + 26 + 34

   0 **D** 34 – 26 + 18

Practice 1.B2

1. Which names the same number as 4 + 31 + 92?

    0 **A** 92 – 31 + 4

    0 **B** 4 x 31 + 92

    0 **C** 92 + 4 – 31

    0 **D** 92 + 4 + 31

2. Which number would make this number sentence correct?

    101 + 43 + ☐ = 43 + 19 + 101

    0 **A** 19

    0 **B** 52

    0 **C** 120

    0 **D** 144

3. Which names the same number as 209 + 126 + 353?

    0 **A** 209 – 126 + 353

    0 **B** 126 x 209 – 353

    0 **C** 353 + 209 + 126

    0 **D** 353 + 126 – 209

4. Which number would make this number sentence correct?

    40 + ☐ + 4 = 400 + 4 + 40

    0 **A** 444

    0 **B** 440

    0 **C** 400

    0 **D** 44

5. Which names the same number as 41 + 28 + 56?

    0 **A** 41 x 28 + 56

    0 **B** 41 + 56 – 28

    0 **C** 56 + 28 – 41

    0 **D** 28 + 56 + 41

6. Which number would make this number sentence correct?

    25 + 45 + 70 = 45 + ☐ + 70

    0 **A** 130

    0 **B** 115

    0 **C** 25

    0 **D** 20

Practice 1.B3

I.B   Identify the number sentence that shows the
       commutative property of addition

1. Which names the same number as 53 + 42 + 64?

   0 **A** 53 – 42 + 64

   0 **B** 42 + 64 + 53

   0 **C** 42 + 64 – 53

   0 **D** 64 + 53 – 42

2. Which number would make this number sentence correct?

   81 + 23 + 72 = 23 + 72 + ☐

   0 **A** 176

   0 **B** 95

   0 **C** 81

   0 **D** 72

3. Which names the same number as 318 + 85 + 93?

   0 **A** 318 – 85 – 93

   0 **B** 318 + 93 – 85

   0 **C** 93 – 85 + 318

   0 **D** 85 + 93 + 318

4. A pet store owner poured 8 pounds of dog food in bin one, 5 pounds in bin two, and 6 pounds in bin three. Which one shows how many pounds of pet food the store owner put in the bins?

   0 **A** 8 – 6 + 5

   0 **B** 8 x 5 – 6

   0 **C** 6 + 5 – 8

   0 **D** 6 + 8 + 5

5. Which number would make this number sentence correct?

   300 + 30 + 3 = 30 + 3 + ☐

   0 **A** 3,000

   0 **B** 333

   0 **C** 300

   0 **D** 33

6. Which names the same number as 116 + 234 + 367?

   0 **A** 367 – 234 – 116

   0 **B** 367 + 116 + 234

   0 **C** 234 + 367 – 116

   0 **D** 367 + 116 – 234

Practice 1.B4

I.B *Identify the number sentence that shows the commutative property of addition*

1. Which names the same number as $34 + 29 + 73$?

   0 **A** $73 - 29 + 34$

   0 **B** $73 - 34 - 29$

   0 **C** $43 + 92 + 37$

   0 **D** $73 + 29 + 34$

2. Which number would make this number sentence correct?

   $49 + 37 + 81 = 37 + 81 + \square$

   0 **A** 37

   0 **B** 49

   0 **C** 118

   0 **D** 167

3. Which names the same number as $201 + 68 + 83$?

   0 **A** $201 - 68 - 83$

   0 **B** $210 + 83 - 68$

   0 **C** $83 - 68 + 201$

   0 **D** $68 + 201 + 83$

4. Mrs. Myer poured 6 ounces of milk in cup one, 12 ounces in cup two, and 9 ounces in cup three. Which one shows how many ounces of milk Mrs. Myer put in the cups?

   0 **A** $12 - 9 + 6$

   0 **B** $6 \times 9 - 12$

   0 **C** $12 + 9 - 6$

   0 **D** $9 + 6 + 12$

5. Which number would make this number sentence correct?

   $500 + 50 + 5 = 5 + 500 + \square$

   0 **A** 50

   0 **B** 555

   0 **C** 500

   0 **D** 5,000

6. Which names the same number as $218 + 342 + 571$?

   0 **A** $571 - 218 - 342$

   0 **B** $342 + 571 + 218$

   0 **C** $342 - 218 + 571$

   0 **D** $342 + 571 - 218$

Practice 1.C1

I.C Add simple fractions

1. $\frac{1}{8} + \frac{2}{8} =$

   0 A $\frac{3}{16}$ 0 C $\frac{2}{8}$

   0 B $\frac{3}{8}$ 0 D $\frac{1}{16}$

2. $\frac{1}{5} + \frac{3}{5} =$

   0 A $\frac{2}{5}$ 0 C $\frac{4}{5}$

   0 B $\frac{3}{10}$ 0 D $\frac{4}{10}$

3. A candy bar had 10 equal sections. Marty ate 2 sections. Tania ate 1 section. How much of the candy bar did Marty and Tania eat in all?

   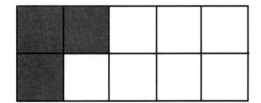

   0 A $\frac{7}{10}$ 0 C $\frac{3}{10}$

   0 B $\frac{3}{5}$ 0 D $\frac{1}{10}$

4. On a field trip, $\frac{1}{4}$ of the class rode in a van and $\frac{2}{4}$ of the class rode in cars. What fraction of the class rode in either a van or a car?

   0 A $\frac{1}{4}$

   0 B $\frac{3}{8}$

   0 C $\frac{2}{4}$

   0 D $\frac{3}{4}$

5. In a game, Debbie won 8 chips. Then she lost 2 chips to Danny and 3 chips to Andrea. What fraction of her chips did Debbie lose?

   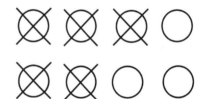

   0 A $\frac{1}{8}$

   0 B $\frac{5}{8}$

   0 C $\frac{5}{16}$

   0 D $\frac{6}{8}$

Practice 1.C2

1. $\frac{2}{7} + \frac{1}{7} =$

    0 **A** $\frac{2}{14}$ 0 **C** $\frac{3}{7}$

    0 **B** $\frac{3}{14}$ 0 **D** $\frac{5}{7}$

2. $\frac{3}{10} + \frac{4}{10} =$

    0 **A** $\frac{7}{20}$ 0 **C** $\frac{3}{4}$

    0 **B** $\frac{7}{10}$ 0 **D** $\frac{9}{10}$

3. Toby is making cookies. For each batch of cookies, he needs $\frac{1}{5}$ cup of honey. How much honey will he need for 3 batches of cookies?

    0 **A** $\frac{3}{15}$ cup

    0 **B** $\frac{3}{10}$ cup

    0 **C** $\frac{2}{5}$ cup

    0 **D** $\frac{3}{5}$ cup

4. Sarah uses $\frac{1}{8}$ of her pay for food and $\frac{2}{8}$ of her pay for rent. What fraction of her money does Sarah use for food and rent?

    0 **A** $\frac{2}{8}$

    0 **B** $\frac{3}{16}$

    0 **C** $\frac{3}{8}$

    0 **D** $\frac{4}{16}$

5. Joshua had 9 trading coins. There was a white star on $\frac{4}{9}$ of his coins and a black star on $\frac{1}{9}$ of his coins. What fraction of his coins had stars?

    0 **A** $\frac{5}{18}$

    0 **B** $\frac{5}{9}$

    0 **C** $\frac{4}{9}$

    0 **D** $\frac{3}{9}$

Practice 1.C3

I.C Add simple fractions

1. $\frac{1}{6} + \frac{4}{6} =$

    O A $\frac{3}{12}$ O C $\frac{3}{6}$

    O B $\frac{4}{12}$ O D $\frac{5}{6}$

2. $\frac{4}{9} + \frac{3}{9} =$

    O A $\frac{1}{9}$ O C $\frac{7}{9}$

    O B $\frac{7}{18}$ O D $\frac{8}{9}$

3. Each morning, Donna uses $\frac{1}{4}$ cup milk on her cereal. How much milk will she use in 3 days?

    O A $\frac{3}{4}$ cup

    O B $\frac{2}{4}$ cup

    O C $\frac{3}{8}$ cup

    O D $\frac{3}{16}$ cup

4. A game board has 10 equal sections. There is a triangle on $\frac{1}{10}$ of the sections and a circle on $\frac{2}{10}$ of the sections. What fraction of the board has a triangle or circle?

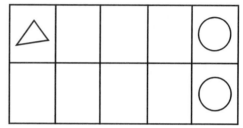

    O A $\frac{3}{20}$

    O B $\frac{3}{5}$

    O C $\frac{3}{10}$

    O D $\frac{7}{10}$

5. Andrew knew that one lap on the track was $\frac{1}{7}$ mile. How far did he walk if he went around the track 3 times?

    O A 1 mile

    O B $\frac{6}{7}$ mile

    O C $\frac{3}{7}$ mile

    O D $\frac{3}{14}$ mile

Practice 1.C4

I.C Add simple fractions

1. $\dfrac{2}{5} + \dfrac{2}{5} =$

    0 **A** $\dfrac{4}{10}$ 0 **C** $\dfrac{2}{10}$

    0 **B** $\dfrac{4}{5}$ 0 **D** $\dfrac{3}{5}$

2. $\dfrac{1}{9} + \dfrac{6}{9} =$

    0 **A** $\dfrac{7}{18}$ 0 **C** $\dfrac{7}{9}$

    0 **B** $\dfrac{5}{9}$ 0 **D** $\dfrac{4}{9}$

3. Raju has two cats. He feeds the larger cat $\dfrac{3}{8}$ cup of food each morning. He feeds the smaller cat $\dfrac{2}{8}$ cup of food. How much food does he feed both cats?

    0 **A** $\dfrac{1}{8}$ cup

    0 **B** $\dfrac{2}{8}$ cup

    0 **C** $\dfrac{5}{8}$ cup

    0 **D** $\dfrac{5}{16}$ cup

4. A loaf of bread was cut into 7 equal slices. Deena ate 2 slices and Mari ate 3 slices. How much of the bread did they eat?

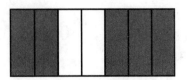

    0 **A** $\dfrac{5}{7}$

    0 **B** $\dfrac{1}{7}$

    0 **C** $\dfrac{2}{7}$

    0 **D** $\dfrac{4}{7}$

5. Denice earned 9 coupons from her teacher. She used 3 coupons to buy a yo-yo and 2 coupons to buy bubblegum. What fraction of her coupons did Denice use?

    0 **A** $\dfrac{2}{3}$

    0 **B** $\dfrac{2}{9}$

    0 **C** $\dfrac{3}{9}$

    0 **D** $\dfrac{5}{9}$

Practice 1.D1

I.D   Add money with and without models

1. Paul spent 62¢ for a bag of chips and 29¢ for a pack of gum. How much did Paul spend?

   0 **A** 47¢

   0 **B** 81¢

   0 **C** 87¢

   0 **D** 91¢

2. Maria had 37¢. Her brother gave her the coins below. How much money did Maria have in all?

   0 **A** 59¢

   0 **B** 69¢

   0 **C** 77¢

   0 **D** 79¢

3. At school, a carton of milk costs 47¢. How much would you spend for two cartons of milk?

   0 **A** 80¢

   0 **B** 84¢

   0 **C** 94¢

   0 **D** $1.04

4. How much money is in the two boxes?

   0 **A** 51¢

   0 **B** 41¢

   0 **C** 31¢

   0 **D** 26

27

Practice 1.D2

I.D Add money with and without models

1. Mr. Moore told each student to bring 39¢ for a new pen and 37¢ for art paper. How much money did each student need?

0 **A** 62¢

0 **B** 66¢

0 **C** 72¢

0 **D** 76¢

2. Chad had 56¢. He found the coins below in his bank. How much money did he have in all?

0 **A** $1.21

0 **B** $1.11

0 **C** $1.01

0 **D** 91¢

3. Mrs. Lee bought a newspaper for 55¢ and a birthday card for $1.25. How much did she spend in all?

0 **A** 70¢

0 **B** $1.70

0 **C** $1.75

0 **D** $1.80

4. How much money is in the two boxes?

0 **A** $1.02

0 **B** $1.07

0 **C** $1.15

0 **D** $1.25

Practice 1.D3

I.D   Add money with and without models

1. 94¢ + 22¢ =

   0 **A** 72¢

   0 **B** $1.06

   0 **C** $1.12

   0 **D** $1.16

2. 58¢ + 33¢ =

   0 **A** 93¢

   0 **B** 91¢

   0 **C** 85¢

   0 **D** 81¢

3. At the Taco House, a taco costs 48¢. How much would you pay for two tacos?

   0 **A** 80¢

   0 **B** 86¢

   0 **C** 96¢

   0 **D** $1.16

4. How much money is in the two boxes?

   0 **A** 75¢

   0 **B** 51¢

   0 **C** 46¢

   0 **D** 41¢

5. At the school store, one pencil costs 17¢. How much would you pay for two pencils?

   0 **A** 36¢

   0 **B** 34¢

   0 **C** 24¢

   0 **D** 30¢

Practice 1.D4

I.D   Add money with and without models

1. 75¢ + 16¢ =

   0 **A** 93¢

   0 **B** 91¢

   0 **C** 81¢

   0 **D** 61¢

2. 27¢ + 56¢ =

   0 **A** 71¢

   0 **B** 73¢

   0 **C** 82¢

   0 **D** 83¢

3. At a sandwich shop, a bag of chips costs 29¢. How much would you pay for two bags of chips?

   0 **A** 58¢

   0 **B** 48¢

   0 **C** 56¢

   0 **D** 59¢

4. How much money is in the two boxes?

   0 **A** 88¢

   0 **B** 36¢

   0 **C** $1.00

   0 **D** 76¢

5. At the school fair, a ride on the merry-go-round costs 35¢. A ride on the roller coaster costs 45¢. How much would it cost to ride both rides?

   0 **A** 70¢

   0 **B** 65¢

   0 **C** 80¢

   0 **D** 85¢

Using Subtraction to Solve Problems

II. Use the operation of subtraction to solve problems

A. Subtract or model subtraction using pictures, words, and numbers
B. Subtract simple fractions
C. Add money with and without models

31

Notes

Objective 2: Pretest

II.A Subtract or model subtraction using pictures, words, and numbers (1-5)

1. Lizette made 9 notecards. She mailed 3 of the cards to her friends. Which one shows how many cards Lizette had then?

0 **A**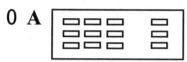

0 **B**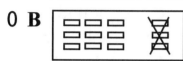

0 **C**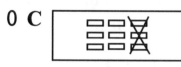

0 **D**

2.   4 0
      − 2 1
      ——

0 **A** 19

0 **B** 21

0 **C** 29

0 **D** 61

3.   6 1 7
      − 4 8
      ——

0 **A** 669

0 **B** 631

0 **C** 569

0 **D** 539

4. Seth made a necklace for his mother. He began with 118 beads. He had 29 beads left at the end. How many beads did he use on the necklace?

0 **A** 111

0 **B** 109

0 **C** 91

0 **D** 89

5. The drive from Diane's home to the beach is 117 miles. If she has driven 58 miles, how much farther must she go?

0 **A** 175 miles

0 **B** 69 miles

0 **C** 61 miles

0 **D** 59 miles

II.B Subtract simple fractions (6-10)

6. $\dfrac{4}{7} - \dfrac{1}{7} =$

    0 **A** $\dfrac{1}{7}$ 0 **C** $\dfrac{5}{7}$

    0 **B** $\dfrac{3}{7}$ 0 **D** $\dfrac{4}{14}$

7. Mrs. Morgan had $\dfrac{4}{5}$ cup of sugar. She used $\dfrac{2}{5}$ cup to make cookies. How much sugar did she have left?

    0 **A** $\dfrac{4}{5}$ cup

    0 **B** $\dfrac{3}{5}$ cup

    0 **C** $\dfrac{2}{5}$ cup

    0 **D** $\dfrac{2}{10}$ cup

8. Mr. Robert needs $\dfrac{5}{8}$ pound of soil for a plant. He has $\dfrac{2}{8}$ pound of soil. How much more does he need?

    0 **A** $\dfrac{7}{8}$ pound

    0 **B** $\dfrac{7}{16}$ pound

    0 **C** $\dfrac{4}{8}$ pound

    0 **D** $\dfrac{3}{8}$ pound

9. $\dfrac{9}{10} - \dfrac{6}{10} =$

    0 **A** $\dfrac{1}{10}$ 0 **C** $\dfrac{3}{10}$

    0 **B** $\dfrac{2}{10}$ 0 **D** $\dfrac{3}{4}$

10. A bag held $\dfrac{7}{8}$ pound of candy. Jenny took $\dfrac{4}{8}$ pound of candy from the bag. How much candy was left in the bag?

    0 **A** $\dfrac{7}{8}$ pound

    0 **B** $\dfrac{3}{16}$ pound

    0 **C** $\dfrac{1}{4}$ pound

    0 **D** $\dfrac{3}{8}$ pound

I.C Subtract money with and without models (11-15)

11. Charles had 75¢ in his pocket. He spent 47¢ for a pack of gum. How much money did he have left?

    0 **A** 28¢

    0 **B** 32¢

    0 **C** 38¢

    0 **D** 44¢

12. Danny has these coins.

How much will he have if he gives away 12¢?

0 **A** 49¢

0 **B** 45¢

0 **C** 35¢

0 **D** 25¢

13. John bought a notebook for 58¢. He gave these coins to the store clerk.

How much did he get back in change?

0 **A** 27¢

0 **B** 23¢

0 **C** 17¢

0 **D** 15¢

14. Adam has these coins.

If he gives his sister 16¢, how much will he have left?

0 **A** 12¢

0 **B** 15¢

0 **C** 25¢

0 **D** 47¢

15. 92¢ – 54¢ =

0 **A** 48¢

0 **B** 46¢

0 **C** 42¢

0 **D** 38¢

Practice 2.A1

II.A Subtract or model subtraction using pictures, words, and numbers

1. There are 12 cookies on a plate. Your brother eats 3 cookies. Which one shows how many cookies would be left?

0 **A**

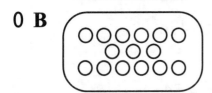

0 **B**

0 **C**

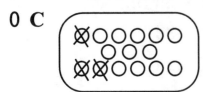

0 **D**

2.   6 2
    − 3 7
    _____

0 **A** 25

0 **B** 26

0 **C** 35

0 **D** 39

3.   7 7 5
    − 5 6
    _____

0 **A** 709

0 **B** 718

0 **C** 719

0 **D** 721

4. This summer, Beth wants to ride 215 miles on her bike. In the first week, she rode 37 miles. How many more miles does she have to ride?

0 **A** 222 miles

0 **B** 188 miles

0 **C** 182 miles

0 **D** 178 miles

5. In May, David's dog weighed 27 pounds. In October, the dog weighed 56 pounds. How many pounds did the dog gain?

0 **A** 83 pounds

0 **B** 39 pounds

0 **C** 31 pounds

0 **D** 29 pounds

Practice 2.A2

II.A Subtract or model subtraction using pictures, words, and numbers

1. Glenn began the game with 25 tokens. He lost 18 during the game. How many tokens did he have then?

$
\begin{array}{ccccc}
\$ & \$ & \$ & \$ & \$ \\
\$ & \$ & \$ & \$ & \$ \\
\$ & \$ & \$ & \$ & \$ \\
\$ & \$ & \$ & \$ & \$ \\
\$ & \$ & \$ & \$ & \$ \\
\end{array}
$

- 0 **A** 43
- 0 **B** 17
- 0 **C** 13
- 0 **D** 7

2.
$$209 - 126$$

- 0 **A** 73
- 0 **B** 83
- 0 **C** 123
- 0 **D** 183

3.
$$815 - 78$$

- 0 **A** 893
- 0 **B** 863
- 0 **C** 767
- 0 **D** 737

4. At nine o'clock, there were 123 cars in a parking lot. At ten o'clock, there were 69 cars in the parking lot. How many cars had left?

- 0 **A** 52
- 0 **B** 54
- 0 **C** 64
- 0 **D** 66

5. Darren had 508 baseball cards. He gave 175 cards to his little brother. How many cards did he have left?

- 0 **A** 473
- 0 **B** 373
- 0 **C** 333
- 0 **D** 323

Practice 2.A3

1. At work, Mrs. Tran sent out the mail. She wanted to mail 14 letters on Monday. She sent 8 in the morning. Which one shows how many more letters she had to send?

0 **A**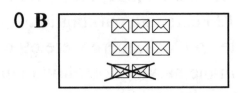

0 **B**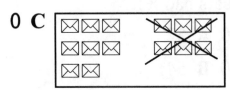

0 **C**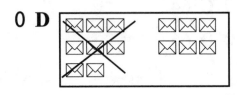

0 **D**

2.   3 1 7
       – 1 2 6
       ———

0 **A** 291

0 **B** 231

0 **C** 211

0 **D** 191

3.   9 4 1
       – 3 6 8
       ———

0 **A** 573

0 **B** 613

0 **C** 623

0 **D** 627

4. On Monday, a baker had 210 pounds of flour. On Friday, he had 75 pounds of flour left. How much flour did he use since Monday?

0 **A** 135 pounds

0 **B** 145 pounds

0 **C** 265 pounds

0 **D** 285 pounds

5. A parking lot can hold 300 cars. There are 64 cars on the lot. How many more cars could park there?

0 **A** 364

0 **B** 246

0 **C** 236

0 **D** 226

Practice 2.A4

II.A Subtract or model subtraction using pictures, words, and numbers

1. Mr. Lee bought a package of 24 star stickers. He gave 17 stickers to his students. How many stickers did he have left?

☆ ☆ ☆ ☆ ☆ ☆ ☆ ☆
☆ ☆ ☆ ☆ ☆ ☆ ☆ ☆
☆ ☆ ☆ ☆ ☆ ☆ ☆ ☆

  0 **A** 7

  0 **B** 17

  0 **C** 27

  0 **D** 41

2.
$$
\begin{array}{r}
3\,0\,3 \\
-\,1\,2\,6 \\
\hline
\end{array}
$$

  0 **A** 177

  0 **B** 187

  0 **C** 223

  0 **D** 277

3.
$$
\begin{array}{r}
5\,2\,2 \\
-\,4\,8\,1 \\
\hline
\end{array}
$$

  0 **A** 141

  0 **B** 51

  0 **C** 41

  0 **D** 161

4. A truck carried 284 boxes of books. It left 196 boxes at the library. How many boxes of books were left on the truck?

  0 **A** 86

  0 **B** 112

  0 **C** 88

  0 **D** 98

5. A theater has 412 seats. At a movie, 198 seats are taken. How many seats are not taken?

  0 **A** 386

  0 **B** 324

  0 **C** 224

  0 **D** 214

Practice 2.B1

II.B   Subtract simple fractions

1. $\frac{3}{5} - \frac{1}{5} =$

   0 A $\frac{2}{10}$ 0 C $\frac{3}{5}$

   0 B $\frac{2}{5}$ 0 D $\frac{4}{5}$

2. Kate had $\frac{3}{4}$ cup of honey. She used $\frac{2}{4}$ cup of honey to make cookies. How much honey did she have left?

   0 A $\frac{4}{5}$ cup

   0 B $\frac{1}{3}$ cup

   0 C $\frac{1}{4}$ cup

   0 D $\frac{1}{8}$ cup

3. Jack needed $\frac{7}{8}$ pound of grass seed. He found $\frac{2}{8}$ pound of grass seed in the shed. How much more grass seed did he need?

   0 A $\frac{9}{16}$ pound

   0 B $\frac{1}{2}$ pound

   0 C $\frac{3}{8}$ pound

   0 D $\frac{5}{8}$ pound

4. $\frac{7}{10} - \frac{4}{10} =$

   0 A $\frac{2}{10}$ 0 C $\frac{7}{10}$

   0 B $\frac{3}{10}$ 0 D $\frac{9}{10}$

5. Donna's bowl held $\frac{5}{6}$ cup of soup. Donna ate $\frac{4}{6}$ cup of soup. How much soup was left in her bowl?

   0 A $\frac{9}{12}$ cup

   0 B $\frac{4}{6}$ cup

   0 C $\frac{2}{6}$ cup

   0 D $\frac{1}{6}$ cup

6. Gary worked in the yard and earned \$6.00. He put $\frac{1}{4}$ of his money in the bank. He spent the rest. What fraction of his money did Gary spend?

   0 A $\frac{3}{4}$

   0 B $\frac{2}{4}$

   0 C $\frac{1}{4}$

   0 D $\frac{1}{6}$

Practice 2.B2

1. $\frac{3}{7} - \frac{1}{7} =$

    0 **A** $\frac{4}{7}$ 0 **C** $\frac{2}{7}$

    0 **B** $\frac{3}{14}$ 0 **D** $\frac{1}{7}$

2. Liz worked on homework for 2 hours. She spent $\frac{1}{6}$ of the time on spelling. What fraction of her time did she spend on other work?

    0 **A** $\frac{1}{8}$

    0 **B** $\frac{1}{2}$

    0 **C** $\frac{4}{6}$

    0 **D** $\frac{5}{6}$

3. A jar holds $\frac{9}{10}$ cup of sugar. Mr. Jackson uses $\frac{8}{10}$ cup of sugar for a cake. How much sugar is left in the jar?

    0 **A** $\frac{1}{10}$ cup

    0 **B** $\frac{2}{10}$ cup

    0 **C** $\frac{7}{10}$ cup

    0 **D** $\frac{9}{10}$ cup

4. $\frac{4}{5} - \frac{2}{5} =$

    0 **A** $\frac{6}{10}$ 0 **C** $\frac{2}{5}$

    0 **B** $\frac{3}{5}$ 0 **D** $\frac{1}{5}$

5. There were 15 guests at a party, and $\frac{3}{5}$ of them were girls. What fraction of the guests were boys?

    0 **A** $\frac{1}{5}$

    0 **B** $\frac{2}{5}$

    0 **C** $\frac{3}{15}$

    0 **D** $\frac{3}{10}$

6. Marsha had $\frac{7}{8}$ yard of string. She used $\frac{2}{8}$ yard of string to make a bracelet. How much string did she have left?

    0 **A** $\frac{1}{8}$ yard

    0 **B** $\frac{2}{8}$ yard

    0 **C** $\frac{3}{8}$ yard

    0 **D** $\frac{5}{8}$ yard

Practice 2.B3

II.B Subtract simple fractions

1. $\frac{5}{6} - \frac{4}{6} =$

  ○ A $\frac{1}{12}$ ○ C $\frac{2}{6}$

  ○ B $\frac{1}{6}$ ○ D $\frac{3}{6}$

2. A cup holds $\frac{7}{8}$ cup of water. If you drink $\frac{5}{8}$ cup of water, how much will be left in the cup?

  ○ A $\frac{2}{16}$ cup

  ○ B $\frac{2}{8}$ cup

  ○ C $\frac{3}{8}$ cup

  ○ D $\frac{6}{8}$ cup

3. A car needs $\frac{4}{5}$ quart of oil. Bill pours in $\frac{2}{5}$ quart of oil. How much more oil does the car need?

  ○ A $\frac{6}{10}$ quart

  ○ B $\frac{3}{5}$ quart

  ○ C $\frac{2}{5}$ quart

  ○ D $\frac{1}{10}$ quart

4. $\frac{7}{10} - \frac{6}{10} =$

  ○ A $\frac{1}{10}$ ○ C $\frac{1}{6}$

  ○ B $\frac{1}{7}$ ○ D $\frac{3}{10}$

5. There were 20 fish in a bowl, and $\frac{3}{4}$ of them were orange. What fraction of the fish were not orange?

  ○ A $\frac{1}{2}$

  ○ B $\frac{1}{3}$

  ○ C $\frac{1}{4}$

  ○ D $\frac{1}{8}$

6. Deepa needs $\frac{6}{7}$ yard of cloth to make a belt. She has $\frac{3}{7}$ yard of cloth. How much more cloth does she need?

  ○ A $\frac{9}{14}$ yard

  ○ B $\frac{4}{7}$ yard

  ○ C $\frac{3}{7}$ yard

  ○ D $\frac{2}{7}$ yard

Practice 2.B4

II.B Subtract simple fractions

1. $\frac{4}{7} - \frac{2}{7} =$

  0 **A** $\frac{4}{7}$ 0 **C** $\frac{2}{14}$

  0 **B** $\frac{2}{7}$ 0 **D** $\frac{6}{7}$

2. A glass held $\frac{3}{5}$ cup of juice. Julie drank $\frac{1}{5}$ cup of juice. How much juice was left in the glass?

  0 **A** $\frac{2}{5}$ cup

  0 **B** $\frac{1}{5}$ cup

  0 **C** $\frac{4}{5}$ cup

  0 **D** $\frac{1}{10}$ cup

3. A bag holds $\frac{3}{8}$ pound of flour. Mrs. George pours $\frac{2}{8}$ pound of flour from the bag. How much flour is left in the bag?

  0 **A** $\frac{2}{8}$ pound

  0 **B** $\frac{5}{16}$ pound

  0 **C** $\frac{5}{8}$ pound

  0 **D** $\frac{1}{8}$ pound

4. $\frac{5}{6} - \frac{4}{6} =$

  0 **A** $\frac{9}{12}$ 0 **C** $\frac{1}{12}$

  0 **B** $\frac{2}{6}$ 0 **D** $\frac{1}{6}$

5. Tandy bought $\frac{7}{8}$ yard of lace. She used $\frac{2}{8}$ yard on an apron. How much lace did she have left?

  0 **A** $\frac{14}{16}$ yard

  0 **B** $\frac{4}{8}$ yard

  0 **C** $\frac{5}{8}$ yard

  0 **D** $\frac{3}{8}$ yard

6. Bobby walked $\frac{9}{10}$ mile in the morning. He walked $\frac{6}{10}$ mile in the afternoon. How much farther did he walk in the morning than in the afternoon?

  0 **A** $\frac{5}{10}$ mile

  0 **B** $\frac{3}{10}$ mile

  0 **C** $\frac{7}{10}$ mile

  0 **D** $\frac{4}{10}$ mile

43

Practice 2.C1

II.C Subtract money with and without models

1. Jeannie had 78¢. She spent 59¢ on gum. How much money did she have left?

0 **A** 17¢

0 **B** 19¢

0 **C** 21¢

0 **D** 29¢

2. Danny had these coins in his pocket.

He spends 17¢ on candy. How much money does he have left?

0 **A** 59¢

0 **B** 35¢

0 **C** 27¢

0 **D** 25¢

3. Mrs. Clark has these coins.

She buys a newspaper for 25¢. How much money does she have left?

0 **A** 3¢

0 **B** 13¢

0 **C** 28¢

0 **D** 53¢

4. 65¢ – 27¢ =

0 **A** 48¢

0 **B** 42¢

0 **C** 38¢

0 **D** 36¢

5. 76¢ – 37¢ =

0 **A** 39¢

0 **B** 41¢

0 **C** 43¢

0 **D** 49¢

Practice 2.C2

II.C Subtract money with and without models

1. Kathy gave the store clerk 75¢ for a soda. The clerk gave Kathy 16¢ in change. How much did the soda cost?

0 **A** 51¢

0 **B** 59¢

0 **C** 61¢

0 **D** 91¢

2. Steve had these coins in his bank.

He gave his little sister 15¢. How much did he have left?

0 **A** 56¢

0 **B** 34¢

0 **C** 26¢

0 **D** 24¢

3. Mr. Adams has these coins in his pocket.

He pays 15¢ for a piece of candy. How much money does he have left?

0 **A** 13¢

0 **B** 23¢

0 **C** 43¢

0 **D** 53¢

4. 65¢ – 27¢ =

0 **A** 48¢

0 **B** 42¢

0 **C** 39¢

0 **D** 38¢

5. 60¢ – 35¢ =

0 **A** 25¢

0 **B** 35¢

0 **C** 45¢

0 **D** 95¢

Practice 2.C3

1. You give the store clerk $1.00 for a 59¢ drink. How much money should the clerk give back to you?

0 **A** 59¢

0 **B** 51¢

0 **C** 41¢

0 **D** 31¢

2. Andy had these coins.

He bought a comic book for 39¢. How much did he have left?

0 **A** 48¢

0 **B** 43¢

0 **C** 42¢

0 **D** 36¢

3. Tyler bought a bag of chips for 69¢. He gave the clerk these coins.

How much money did he get in change?

0 **A** 20¢

0 **B** 14¢

0 **C** 8¢

0 **D** 6¢

4. 62¢ – 34¢ =

0 **A** 28¢

0 **B** 32¢

0 **C** 36¢

0 **D** 38¢

5. 80¢ – 59¢ =

0 **A** 19¢

0 **B** 21¢

0 **C** 31¢

0 **D** 39¢

Using Multiplication to Solve Problems

III. Use the operation of multiplication to solve problems

A. Solve and record multiplication problems (one-digit multipliers)

Notes

Objective 3: Pretest

III.A Solve and record multiplication problems [one-digit multipliers] (1-18)

1. Tanya gave a bag of candy to her 3 friends. Each bag had 5 pieces of candy. How many pieces of candy did Tanya give to her friends?

  0 **A** 2

  0 **B** 8

  0 **C** 12

  0 **D** 15

2. Mr. Wayne went to the store 4 times. He used 4 coupons on each trip to the store. How many coupons did he use in all?

  0 **A** 8

  0 **B** 12

  0 **C** 16

  0 **D** 20

3. Each pizza from Pizza Palace is cut into 8 pieces. How many pieces would there be in 3 pizzas?

  0 **A** 24

  0 **B** 22

  0 **C** 11

  0 **D** 5

4. David can read 10 pages in one hour. How many pages can he read in 3 hours?

  0 **A** 10

  0 **B** 13

  0 **C** 30

  0 **D** 20

5. Rebecca bought 6 packs of pencils. There were 3 pencils in each pack. How many pencils did she buy in all?

  0 **A** 2

  0 **B** 9

  0 **C** 16

  0 **D** 18

6. A clerk earns $5.00 per hour. How much would the clerk earn in 4 hours?

  0 **A** $10.00

  0 **B** $15.00

  0 **C** $19.00

  0 **D** $20.00

7.   8

   x 7

0 **A** 64

0 **B** 56

0 **C** 48

0 **D** 15 ✓

8.   9

   x 3

0 **A** 12

0 **B** 18

0 **C** 27

0 **D** 33 ✓

9.   7

   x 4

0 **A** 11

0 **B** 14

0 **C** 21

0 **D** 28 ✓

10. Gloria walks 12 miles each week. How many miles does she walk in 3 weeks?

0 **A** 15

0 **B** 18

0 **C** 24

0 **D** 36 ✓

11. A pitcher holds 14 cups. How many cups would 3 pitchers hold?

0 **A** 48

0 **B** 42

0 **C** 32

0 **D** 17 ✓

12. A school van holds 15 people. How many people would 3 vans hold?

0 **A** 18

0 **B** 35

0 **C** 45

0 **D** 50 ✓

13. There are 24 children in each third-grade class. There are 4 third-grade classes. How many children are in the third grade?

  0 **A** 28

  0 **B** 84

  0 **C** 86

  0 **D** 96

14. At the store, apples are kept in crates. Each crate holds 14 apples. If you buy 5 crates, how many apples will you get?

  0 **A** 70

  0 **B** 60

  0 **C** 50

  0 **D** 20

15. Maria baked 6 batches of cookies. Each batch of cookies made 12 cookies. How many cookies did Maria bake?

  0 **A** 18

  0 **B** 48

  0 **C** 62

  0 **D** 72

16.
$$\begin{array}{r} 1\,7 \\ \times\,4 \\ \hline \end{array}$$

  0 **A** 68

  0 **B** 58

  0 **C** 48

  0 **D** 46

17.
$$\begin{array}{r} 2\,1 \\ \times\,4 \\ \hline \end{array}$$

  0 **A** 104

  0 **B** 94

  0 **C** 84

  0 **D** 76

18.
$$\begin{array}{r} 1\,8 \\ \times\,6 \\ \hline \end{array}$$

  0 **A** 94

  0 **B** 98

  0 **C** 106

  0 **D** 108

Practice 3.A1

III.A Solve and record multiplication problems (one-digit multipliers)

1. Kiran bought 4 folders. Each folder had 8 pockets. How many pockets did the folders have in all?

    0 **A** 16

    0 **B** 24

    0 **C** 32

    0 **D** 36

2. A can of soda holds 8 ounces. How many ounces are in 5 cans of soda?

    0 **A** 24

    0 **B** 40

    0 **C** 45

    0 **D** 50

3.   9
       x 4
       ——

    0 **A** 18

    0 **B** 27

    0 **C** 36

    0 **D** 45

4. A car can go 23 miles on one gallon of gas. How many miles can the car go on 4 gallons of gas?

    0 **A** 82

    0 **B** 87

    0 **C** 92

    0 **D** 96

5. Mrs. Martinez made party bags for each child at a party. She put 14 pieces of candy in each bag. She made 7 bags. How many pieces of candy did she use?

    0 **A** 81

    0 **B** 88

    0 **C** 91

    0 **D** 98

6.   2 4
       x 3
       ——

    0 **A** 74

    0 **B** 72

    0 **C** 67

    0 **D** 62

Practice 3.A2

III.A Solve and record multiplication problems (one-digit multipliers)

1. Each box holds 8 pencils. How many pencils would there be in 6 boxes?

   0 **A** 14

   0 **B** 32

   0 **C** 42

   0 **D** 48

2. Heidi walks 6 miles each week. How many miles will she walk in 6 weeks?

   0 **A** 36

   0 **B** 24

   0 **C** 18

   0 **D** 12

3.　　8
   　×7

   0 **A** 15

   0 **B** 48

   0 **C** 56

   0 **D** 64

4. Mia's dogs eat 16 pounds of dog food each week. How many pounds of dog food will they eat in 4 weeks?

   0 **A** 44

   0 **B** 54

   0 **C** 64

   0 **D** 68

5. Tom bought 5 shirts. Each shirt cost $17. How much did Tom pay for all the shirts?

   0 **A** $85

   0 **B** $75

   0 **C** $72

   0 **D** $52

6.　　2 6
   　×3

   0 **A** 68

   0 **B** 69

   0 **C** 78

   0 **D** 84

Practice 3.A3

III.A Solve and record multiplication problems
(one-digit multipliers)

1. Inez made 7 cakes for the school fair. She used 3 cups of sugar in each cake. How many cups of sugar did she use in all?

   0 **A** 10

   0 **B** 21

   0 **C** 24

   0 **D** 28

2. Mrs. Perez made 6 flower boxes. She put 9 plants in each box. How many plants did she use?

   0 **A** 42

   0 **B** 46

   0 **C** 54

   0 **D** 56

3. 7 x 7 =

   0 **A** 56

   0 **B** 52

   0 **C** 49

   0 **D** 42

4. Sam bought 12 cases of paper. Each case had 8 packages of paper. How many packages of paper did Sam buy?

   0 **A** 20

   0 **B** 72

   0 **C** 84

   0 **D** 96

5. Nick bought 4 boxes of candy. Each box cost $14. How much did Nick pay for the candy?

   0 **A** $42

   0 **B** $48

   0 **C** $56

   0 **D** $66

6.
   $$\begin{array}{r} 19 \\ \times 4 \\ \hline \end{array}$$

   0 **A** 42

   0 **B** 54

   0 **C** 66

   0 **D** 76

Practice 3.A4

1. Greg has 8 boxes of crayons. Each box has 8 crayons. How many crayons does Greg have in all?

   0 **A** 16

   0 **B** 32

   0 **C** 48

   0 **D** 64

2. Mr. Reeves drinks 5 cups of coffee a day. How many cups of coffee does he drink in 7 days?

   0 **A** 35

   0 **B** 30

   0 **C** 25

   0 **D** 12

3. 9 x 8 =

   0 **A** 58

   0 **B** 66

   0 **C** 72

   0 **D** 84

4. Mary can read 22 words in a minute. How many words can she read in 5 minutes?

   0 **A** 94

   0 **B** 107

   0 **C** 110

   0 **D** 120

5. Alberto drinks 15 bottles of water each week. How many bottles of water will he drink in 5 weeks?

   0 **A** 50

   0 **B** 55

   0 **C** 65

   0 **D** 75

6. $$\begin{array}{r} 3\,2 \\ \times\ 5 \\ \hline \end{array}$$

   0 **A** 180

   0 **B** 160

   0 **C** 150

   0 **D** 130

Notes

Using Division to Solve Problems

IV. Use the operation of division to solve problems

A. Use models to solve simple division problems
B. Solve division problems with multi-digit dividend (one-digit divisor)
C. Determine unit cost when given total cost and number of units

Notes

Objective 4: Pretest

*IV.A Use models to solve simple division problems
(1-3)*

1. Stacy has 15 pictures. Which one shows how she could divide the pictures evenly on 3 pages of a scrapbook?

0 **A**

0 **B**

0 **C**

0 **D**

2. Mrs. Chacko had 8 pieces of candy. She divided the candy equally among her 4 children. Which one shows how Mrs. Chacko divided the candy?

0 **A**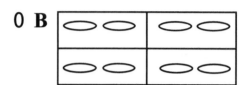

0 **B**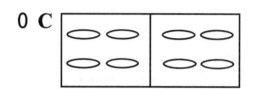

0 **C**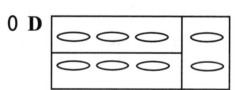

0 **D**

3. Which number sentence matches this picture?

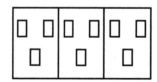

0 **A** 9 x 3 = 27

0 **B** 3 ÷ 3 = 1

0 **C** 12 ÷ 4 = 3

0 **D** 9 ÷ 3 = 3

*IV.B Solve division problems with multi-digit
dividend and one-digit divisor (4-9)*

4. Sherry had 16 beads. She made
4 necklaces and put the same
number of beads on each one.
How many beads were on each
necklace?

0 **A** 12

0 **B** 6

0 **C** 4

0 **D** 3

5. A bag holds 84 pieces of
candy. Danny divided the
candy equally into 2 bowls.
How many pieces of candy did
he put in each bowl?

0 **A** 38

0 **B** 42

0 **C** 44

0 **D** 86

6. A band has 48 members. The
members form 4 equal lines.
How many members are in
each line?

0 **A** 52

0 **B** 42

0 **C** 22

0 **D** 12

7. $96 \div 3 =$

0 **A** 32

0 **B** 33

0 **C** 42

0 **D** 99

8. $86 \div 2 =$

0 **A** 23

0 **B** 32

0 **C** 42

0 **D** 43

9. $57 \div 3 =$

0 **A** 19

0 **B** 23

0 **C** 29

0 **D** 60

©ECS Learning Systems, Inc.

IV.C *Determine unit cost when given total cost and number of units (10-15)*

10. Therese bought 6 blue folders for 48¢. How much did each folder cost?

0 **A** 52¢

0 **B** 16¢

0 **C** 9¢

0 **D** 8¢

11. A box of candy bars costs 52¢. The box has 4 candy bars. How much does each candy bar cost?

0 **A** $2.08

0 **B** $1.00

0 **C** 13¢

0 **D** 10¢

12. A bag holds 5 cups of beans and costs 65¢. What is the cost of one cup of beans?

0 **A** 13¢

0 **B** 15¢

0 **C** 75¢

0 **D** $3.00

13. The cost for 3 white shirts is $75. How much does each shirt cost?

0 **A** $78

0 **B** $27

0 **C** $25

0 **D** $21

14. A box of pencils costs 96¢. There are 8 pencils in the box. What is the cost of one pencil?

0 **A** 10¢

0 **B** 12¢

0 **C** 20¢

0 **D** 24¢

15. Kevin paid 88¢ for 4 bags of chips. How much did each bag cost?

0 **A** 16¢

0 **B** 20¢

0 **C** 22¢

0 **D** 24¢

Practice 4.A1

IV.A Use models to solve simple division problems

1. There are 12 students in Mrs. Robb's math class. She wants them to work in groups of 4. Which one shows how many groups there will be?

0 **A**

0 **B**

0 **C**

0 **D**

2. Which number sentence matches this picture?

0 **A** $12 \div 6 = 2$

0 **B** $12 \div 4 = 2$

0 **C** $18 \div 3 = 6$

0 **D** $18 \div 9 = 2$

3. Which number sentence matches this picture?

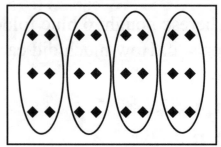

0 **A** $24 \div 4 = 6$

0 **B** $24 \div 12 = 2$

0 **C** $24 \div 8 = 3$

0 **D** $24 \div 2 = 12$

4. Which number sentence matches this picture?

0 **A** $15 \div 3 = 5$

0 **B** $10 \div 5 = 2$

0 **C** $10 \div 1 = 10$

0 **D** $10 \div 10 = 1$

Practice 4.A2

IV.A Use models to solve simple division problems

1. There are 14 students on a baseball team. They will go to their next game on 2 vans. Which one shows how many students can go on each van?

0 **A**

0 **B**

0 **C**

0 **D**

2. Which number sentence matches this picture?

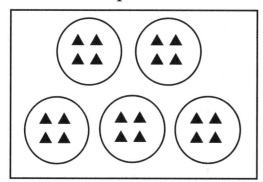

0 **A** $10 \times 2 = 20$

0 **B** $20 - 5 = 15$

0 **C** $20 \div 10 = 2$

0 **D** $20 \div 5 = 4$

3. Which number sentence matches this picture?

0 **A** $12 \div 4 = 3$

0 **B** $16 \div 4 = 4$

0 **C** $16 \div 8 = 2$

0 **D** $16 \div 2 = 8$

4. Which number sentence matches this picture?

0 **A** $24 \div 3 = 8$

0 **B** $28 \div 7 = 4$

0 **C** $35 \div 7 = 5$

0 **D** $21 \div 3 = 7$

Practice 4.A3

IV.A Use models to solve simple division problems

1. Mr. Dawson had 9 new pencils. He gave an equal number of pencils to his 3 sons. Which one shows how many pencils each boy got?

O **A**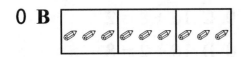

O **B**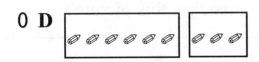

O **C**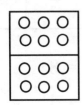

O **D**

2. Which number sentence matches this picture?

O **A** 12 x 2 = 24

O **B** 12 − 2 = 10

O **C** 12 ÷ 3 = 4

O **D** 12 ÷ 2 = 6

3. Which number sentence matches this picture?

O **A** 18 ÷ 9 = 2

O **B** 18 − 6 = 12

O **C** 18 + 3 = 21

O **D** 18 ÷ 3 = 6

4. The picture shows a plate of cookies. Which number sentence matches this picture?

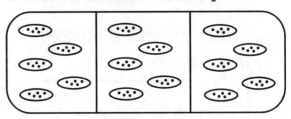

O **A** 5 + 3 = 8

O **B** 15 − 5 = 10

O **C** 15 ÷ 3 = 5

O **D** 18 ÷ 3 = 6

Practice 4.B1

IV.B Solve division problems with multi-digit dividend (one-digit divisor)

1. A box of candy holds 20 pieces. Mary divided the candy equally into 4 bowls. How many pieces of candy did she put in each bowl?

    0 **A** 80

    0 **B** 24

    0 **C** 6

    0 **D** 5

2. There are 25 children and 5 tables in a room. The same number of children sit at each table. How many children sit at each table?

    0 **A** 4

    0 **B** 5

    0 **C** 20

    0 **D** 30

3. Darla planted 28 plants in 7 equal rows. How many plants were in each row?

    0 **A** 35

    0 **B** 21

    0 **C** 6

    0 **D** 4

4. $46 \div 2 =$

    0 **A** 12

    0 **B** 22

    0 **C** 23

    0 **D** 48

5. $63 \div 3 =$

    0 **A** 32

    0 **B** 31

    0 **C** 23

    0 **D** 21

6. $48 \div 4 =$

    0 **A** 12

    0 **B** 13

    0 **C** 22

    0 **D** 44

Practice 4.B2

IV.B Solve division problems with multi-digit dividend (one-digit divisor)

1. Mrs. Terry took 9 children to the playground. There were 3 benches. The same number of children sat on each bench. How many children sat on each bench?

   0 **A** 15

   0 **B** 12

   0 **C** 6

   0 **D** 3

2. Elizabeth has 16 bows. If she puts 2 bows on each package, how many packages can she make?

   0 **A** 7

   0 **B** 8

   0 **C** 9

   0 **D** 18

3. $93 \div 3 =$

   0 **A** 31

   0 **B** 30

   0 **C** 21

   0 **D** 20

4. A package of paper holds 54 sheets of paper. Each student needs 9 pieces of paper for an art project. How many children can work from one package?

   0 **A** 6

   0 **B** 8

   0 **C** 9

   0 **D** 10

5. $84 \div 4 =$

   0 **A** 20

   0 **B** 21

   0 **C** 31

   0 **D** 80

6. $36 \div 2 =$

   0 **A** 12

   0 **B** 13

   0 **C** 18

   0 **D** 28

Practice 4.B3

IV.B Solve division problems with multi-digit dividend (one-digit divisor)

1. Marty had 12 eggs. He needed 2 eggs for a batch of cookies. How many batches of cookies could he make?

   0 **A** 6

   0 **B** 8

   0 **C** 10

   0 **D** 24

2. Toni had 24 flowers. She put 6 flowers in each vase. How many vases did she have?

   0 **A** 4

   0 **B** 6

   0 **C** 18

   0 **D** 30

3. $92 \div 2 =$

   0 **A** 31

   0 **B** 41

   0 **C** 46

   0 **D** 90

4. There are 36 cookies in 4 equal stacks. How many cookies are in each stack?

   0 **A** 40

   0 **B** 10

   0 **C** 9

   0 **D** 8

5. $81 \div 3 =$

   0 **A** 30

   0 **B** 27

   0 **C** 23

   0 **D** 21

6. $56 \div 2 =$

   0 **A** 33

   0 **B** 28

   0 **C** 23

   0 **D** 22

Practice 4.C1

IV.C Determine unit cost when given total cost and number of units

1. A box of soap costs 96¢. There are 3 bars of soap in the box. Each bar of soap costs—

   0 **A** 93¢

   0 **B** 33¢

   0 **C** 32¢

   0 **D** 22¢

2. Tim paid 36¢ for 6 pieces of bubblegum. How much did each piece of bubblegum cost?

   0 **A** 42¢

   0 **B** 30¢

   0 **C** 8¢

   0 **D** 6¢

3. Mr. Ingalls paid $1.00 for 4 newspapers. Each newspaper cost—

   0 **A** 50¢

   0 **B** 25¢

   0 **C** 20¢

   0 **D** 15¢

4. A box of erasers costs 66¢. There are 3 erasers in the box. Each eraser costs—

   0 **A** 20¢

   0 **B** 22¢

   0 **C** 30¢

   0 **D** 33¢

5. At a football game, 5 cups of lemonade cost $1.00. How much did each cup of lemonade cost?

   0 **A** 20¢

   0 **B** 24¢

   0 **C** 25¢

   0 **D** 30¢

6. Kevin paid $2.00 for 4 bags of chips. How much did each bag cost?

   0 **A** $1.00

   0 **B** 75¢

   0 **C** 50¢

   0 **D** 25¢

Practice 4.C2

IV.C Determine unit cost when given total cost and number of units

1. A package of markers cost 60¢. There were 5 markers in the package. Each marker cost—

    0 **A** 10¢

    0 **B** 12¢

    0 **C** 55¢

    0 **D** 65¢

2. Josh paid 45¢ for 5 baseball cards. Each baseball card cost—

    0 **A** 7¢

    0 **B** 8¢

    0 **C** 9¢

    0 **D** 10¢

3. Terry spent 54¢ for pencils. He bought 9 pencils. How much did each pencil cost?

    0 **A** 5¢

    0 **B** 6¢

    0 **C** 8¢

    0 **D** 9¢

4. The cost for 4 baseball caps is $72. Each baseball cap costs—

    0 **A** $18

    0 **B** $19

    0 **C** $22

    0 **D** $23

5. Marsha spent 96¢ to buy 8 dill pickles. Each pickle cost—

    0 **A** 10¢

    0 **B** 11¢

    0 **C** 12¢

    0 **D** 11¢

6. Kevin paid 65¢ for 5 toy cars. How much did each toy car cost?

    0 **A** 10¢

    0 **B** 11¢

    0 **C** 13¢

    0 **D** 15¢

Practice 4.C3

IV.C *Determine unit cost when given total cost and number of units*

1. Mrs. Lee bought 6 small goldfish for 42¢. How much did each fish cost?

   0 **A** 36¢

   0 **B** 9¢

   0 **C** 8¢

   0 **D** 7¢

2. Greta bought 8 sheets of drawing paper for 56¢. How much did each sheet of paper cost?

   0 **A** 12¢

   0 **B** 9¢

   0 **C** 7¢

   0 **D** 6¢

3. Matt bought 3 small gift boxes for 27¢. How much did each gift box cost?

   0 **A** 7¢

   0 **B** 9¢

   0 **C** 24¢

   0 **D** 30¢

4. A box of 5 erasers costs 95¢. How much does each eraser cost?

   0 **A** 15¢

   0 **B** 17¢

   0 **C** 19¢

   0 **D** 21¢

5. Sherry paid 98¢ for 2 Super Comic cards. How much did each card cost?

   0 **A** 34¢

   0 **B** 44¢

   0 **C** 48¢

   0 **D** 49¢

6. Chris spent 90¢ to play a video game 6 times. How much did he spend to play each game?

   0 **A** 14¢

   0 **B** 15¢

   0 **C** 17¢

   0 **D** 20¢

Estimate Solutions

V. Estimate solutions to a problem situation

A. Estimate sums and differences beyond basic facts

Notes

Objective 5: Pretest

V.A   Estimate sums and differences beyond basic facts (1-12)

1. Donald bought a candy bar for 39¢ and a soda for 52¢. About how much did Donald spend in all?

   0 **A** 40¢

   0 **B** 60¢

   0 **C** 80¢

   0 **D** 90¢

2. Donna needs 58 tickets to buy a stuffed animal at the school store. She has 39 tickets. About how many more tickets does Donna need?

   0 **A** 10

   0 **B** 20

   0 **C** 30

   0 **D** 40

3. Gary earned 39 points on field day. Molly earned 52 points. Which would be the best way to find about how many points Gary and Molly earned in all?

   0 **A** 30 + 40

   0 **B** 30 + 50

   0 **C** 40 + 50

   0 **D** 50 + 50

4. There are 79 students who take art classes and 139 students who take music class. About how many more students take music than art?

   0 **A** 40

   0 **B** 60

   0 **C** 80

   0 **D** 220

5. A bag of sugar holds 128 ounces. You use 18 ounces to make cookies. Which would be the best way to find about how much sugar you have left?

   0 **A** 130 – 20

   0 **B** 130 – 10

   0 **C** 120 – 30

   0 **D** 120 – 10

6. Cliff earned $61 for doing yard work. Bill earned $49 for painting houses. About how much more than Bill did Cliff make?

   0 **A** $110

   0 **B** $90

   0 **C** $30

   0 **D** $10

7. In May, Mr. Marzano hiked 82 miles. Mrs. Marzano hiked 119 miles. About how many more miles did Mrs. Marzano hike than Mr. Marzano?

0 **A** 200

0 **B** 60

0 **C** 40

0 **D** 20

8. Team A earned 62 points in a spelling contest. Team B earned 97 points. Which would be the best way to find about how many more points Team B earned than Team A?

0 **A** 100 – 60

0 **B** 100 – 70

0 **C** 90 – 60

0 **D** 90 – 70

9. Brett and his father planted 38 tomato plants and 17 pepper plants. About how many plants did they plant in all?

0 **A** 70

0 **B** 60

0 **C** 40

0 **D** 20

10. Doreen made 18 aprons. Molly made 12 aprons. Which would be the best way to find about how many aprons the girls made in all?

0 **A** 20 + 20

0 **B** 20 + 30

0 **C** 10 + 30

0 **D** 20 + 10

11. Which is the best estimate of 38 + 42?

0 **A** 10

0 **B** 60

0 **C** 70

0 **D** 80

12. Which is the best estimate of 309 – 198?

0 **A** 100

0 **B** 200

0 **C** 300

0 **D** 500

Practice 5.A1

V.A Estimate sums and differences beyond basic facts

1. Sheena's family is driving 138 miles from their home to their aunt's home. They have driven 47 miles. About how many more miles do they still have to drive ?

   0 **A** 110

   0 **B** 100

   0 **C** 90

   0 **D** 70

2. Craig sold 17 boxes of candy. Cindy sold 19 boxes. Which would be the best way to find about how many boxes of candy they sold together?

   0 **A** 10 + 10

   0 **B** 20 + 10

   0 **C** 20 + 20

   0 **D** 20 + 30

3. Which is the best estimate of 289 + 306?

   0 **A** 100

   0 **B** 400

   0 **C** 500

   0 **D** 600

4. On Monday, 78 students ate breakfast at school. On Tuesday, 57 students ate breakfast at school. About how many more students ate breakfast on Monday than on Tuesday?

   0 **A** 10

   0 **B** 20

   0 **C** 40

   0 **D** 140

5. Crystal made 112 cookies for a bake sale. She sold 97 cookies. Which would be the best way to find about how many cookies she did not sell?

   0 **A** 120 – 90

   0 **B** 110 – 110

   0 **C** 120 – 100

   0 **D** 110 – 100

6. Clay and Cara each have 47 baseball cards. About how many do they have in all?

   0 **A** 100

   0 **B** 90

   0 **C** 80

   0 **D** 50

Practice 5.A2

1. Diana needs 576 beads to make necklaces for her friends. She has 298 beads. About how many more beads does she need?

   0 **A** 900

   0 **B** 400

   0 **C** 300

   0 **D** 200

2. A pet store had 123 goldfish. The store sold 51 fish in one day. Which would be the best way to find about how many goldfish the store still had?

   0 **A** 150 – 50

   0 **B** 120 – 50

   0 **C** 110 – 60

   0 **D** 100 – 50

3. Which is the best estimate of 420 – 287?

   0 **A** 100

   0 **B** 200

   0 **C** 300

   0 **D** 700

4. Jason has two dogs, Ike and Mike. In one month, Ike eats 37 pounds of dog food. Mike eats 29 pounds of dog food. About how much food do Ike and Mike eat in one month?

   0 **A** 40 pounds

   0 **B** 50 pounds

   0 **C** 70 pounds

   0 **D** 90 pounds

5. Jenny went bowling. She scored 79 in her first game and 82 in her second game. Which would be the best way to find about how many points she scored in both games?

   0 **A** 80 + 80

   0 **B** 80 + 70

   0 **C** 90 + 80

   0 **D** 100 + 70

6. Rachel and Rebecca each sold 32 boxes of cookies. About how many boxes did they sell together?

   0 **A** 100

   0 **B** 80

   0 **C** 70

   0 **D** 60

Practice 5.A3

V.A *Estimate sums and differences beyond basic facts*

1. Mrs. Leach bought 28 plants for her backyard and 48 plants for her front yard. Which would be the best way to find about how many plants she bought in all?

   0 **A** 30 + 20

   0 **B** 40 + 20

   0 **C** 50 + 20

   0 **D** 50 + 30

2. Mr. Crest drove 489 miles in May and 502 miles in June. About how many miles did he travel during May and June?

   0 **A** 1,000

   0 **B** 900

   0 **C** 800

   0 **D** 700

3. In the summer, Keith and Jay each earned 278 points in a math contest. About how many points did they earn in all?

   0 **A** 400

   0 **B** 500

   0 **C** 600

   0 **D** 700

4. Elise needs 47 inches of lace for a dress and 49 inches of lace for a vest. Which would be the best way to find about how many inches of lace she needs in all?

   0 **A** 40 + 40

   0 **B** 40 + 50

   0 **C** 50 + 50

   0 **D** 50 + 60

5. Trent had $163 at the start of his vacation. At the end, he had $69. Which would be the best way to find about how much money he spent?

   0 **A** $160 – $60

   0 **B** $160 – $70

   0 **C** $200 – $70

   0 **D** $200 – $100

6. Which is the best estimate of 74 + 68?

   0 **A** 200

   0 **B** 170

   0 **C** 160

   0 **D** 140

Practice 5.A4

1. Kerry weighs 58 pounds. Janie weighs 61 pounds. Which is the best estimate of how much the girls weigh together?

   0 **A** 100 pounds

   0 **B** 110 pounds

   0 **C** 120 pounds

   0 **D** 140 pounds

2. In one month, Chris drank 42 cups of milk. Ginny drank 51 cups of milk. About how many cups did the girls drink in all?

   0 **A** 80

   0 **B** 90

   0 **C** 100

   0 **D** 110

3. Which is the best estimate of 508 – 222?

   0 **A** 700

   0 **B** 400

   0 **C** 300

   0 **D** 200

4. John spent $39 at the waterpark. Kent spent $22. Which would be the best way to find about how much more John spent than Kent?

   0 **A** $40 – $20

   0 **B** $40 – $30

   0 **C** $40 – $10

   0 **D** $50 – $20

5. Maria read for 127 minutes on Monday. She read for 52 minutes on Tuesday. About how many more minutes did she read on Monday than on Tuesday?

   0 **A** 50

   0 **B** 60

   0 **C** 70

   0 **D** 80

6. Mr. Hansen bought a new shirt for $29 and new slacks for $61. About how much did he spend in all?

   0 **A** $100

   0 **B** $90

   0 **C** $80

   0 **D** $70

Solution Strategies/ Analyze or Solve Problems

VI. Determine solution strategies and analyze or solve problems

  A. Select and use addition or subtraction to solve problems
  B. Select multiplication (one-digit multiplier) or division (one-digit divisor) and use the operation to solve problems
  C. Measure to solve problems involving length, area, temperature, and time

Notes

Objective 6: Pretest

VI.A Select and use addition or subtraction to solve problems (1-5)

1. The principal needed 36 girls to go on a special field trip. Mrs. Wong chose 12 girls. How many more students did the principal still need?

   0 **A** 48

   0 **B** 44

   0 **C** 28

   0 **D** 24

2. Kristi had 84 beads. She used 27 on a bracelet. How many beads did Kristi have left?

   0 **A** 57

   0 **B** 63

   0 **C** 103

   0 **D** 111

3. Ms. Eng's class collected 246 pounds of paper. Ms. Hoff's class collected 343 pounds. How many pounds did the two classes collect?

   0 **A** 97

   0 **B** 103

   0 **C** 569

   0 **D** 589

4. Jamal earned 67 points in math class. Marcus earned 92 points. How many points did they earn in all?

   0 **A** 25

   0 **B** 155

   0 **C** 159

   0 **D** 169

5. Adele studied for 35 minutes on Monday and for 58 minutes on Tuesday. How many minutes did she study in all?

   0 **A** 103

   0 **B** 93

   0 **C** 83

   0 **D** 23

VI.B Select and use multiplication (one-digit multiplier) or division (one-digit divisor) to solve problems (6-10)

6. Sue has 12 pieces of candy to give to 4 friends. If each friend gets the same amount of candy, how many pieces will each one get?

   0 **A** 48

   0 **B** 36

   0 **C** 4

   0 **D** 3

7. Mrs. Heath ordered 15 boxes of soap. Each box held 3 bars of soap. How many bars of soap did Mrs. Heath order?

0 **A** 45

0 **B** 35

0 **C** 12

0 **D** 3

8. Larry planted 16 plants in 4 equal rows. How many plants were in each row?

0 **A** 64

0 **B** 54

0 **C** 12

0 **D** 4

9. A band had 125 members. If they march in rows of 5, how many rows will there be?

0 **A** 15

0 **B** 25

0 **C** 505

0 **D** 625

10. Frank bought 8 cartons of soda. There were 4 cans in each carton. How many cans of soda did he buy?

0 **A** 2

0 **B** 24

0 **C** 32

0 **D** 36

VI.C Measure to solve problems involving length, area, temperature, and time (11-18)

11. What is the **perimeter** (distance around) this triangle?

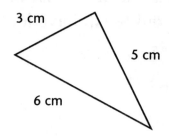

0 **A** 8 cm

0 **B** 11 cm

0 **C** 14 cm

0 **D** 15 cm

12. Tori walked from A to B, and then from B to C. How far did she walk?

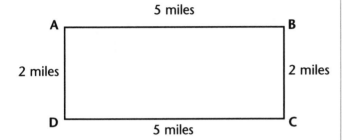

- 0 **A** 10 miles
- 0 **B** 9 miles
- 0 **C** 7 miles
- 0 **D** 4 miles

13. What is the **area** of the rectangle?

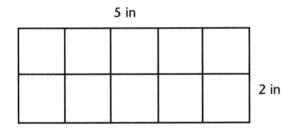

- 0 **A** 2 square inches
- 0 **B** 5 square inches
- 0 **C** 7 square inches
- 0 **D** 10 square inches

14. Look at the triangle inside the rectangle.

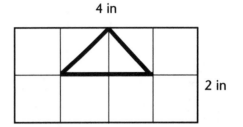

The **area** of the triangle is about—

- 0 **A** 1 square inch
- 0 **B** 2 square inches
- 0 **C** 6 square inches
- 0 **D** 8 square inches

15. What temperature is shown on the thermometer?

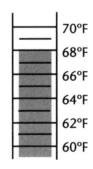

- 0 **A** 70°F
- 0 **B** 68°F
- 0 **C** 66°F
- 0 **D** 64°F

16. Which thermometer shows 36°C?

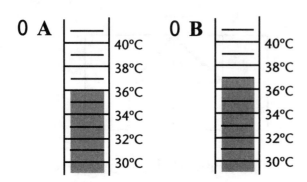

17. What time is shown on this clock?

0 **A** 8:00

0 **B** 7:30

0 **C** 7:00

0 **D** 6:30

18. Look at the clock below. What time was it 20 minutes ago?

0 **A** 9:25

0 **B** 9:00

0 **C** 8:50

0 **D** 8:45

Practice 6.A1

1. On Monday, a baker had 128 eggs. By Wednesday, 49 eggs were left. How many eggs had the baker used?

   0 **A** 177

   0 **B** 89

   0 **C** 81

   0 **D** 79

2. In their first game, a basketball team made 39 points. In their second game, the team made 56 points. How many points did the team make in all?

   0 **A** 105

   0 **B** 85

   0 **C** 95

   0 **D** 17

3. Ben took 408 pounds of paper to the dump. Rob took 279 pounds. How many more pounds did Ben take than Rob?

   0 **A** 129

   0 **B** 131

   0 **C** 239

   0 **D** 687

4. Meg rode 325 miles on a train and 255 miles on a bus. How many miles did she ride in all?

   0 **A** 230

   0 **B** 270

   0 **C** 570

   0 **D** 580

5. A store owner had 84 packs of gum to sell. He bought 96 more packs. How many packs of gum did he have then?

   0 **A** 180

   0 **B** 172

   0 **C** 170

   0 **D** 12

6. Carson School has 348 students. Martin School has 524 students. How many more students are at Martin School than Carson School?

   0 **A** 872

   0 **B** 176

   0 **C** 222

   0 **D** 286

Practice 6.A2

VI.A *Select and use addition or subtraction to solve problems*

1. Mrs. Lyons weighed 142 pounds in May. In July, she weighed 128 pounds. How much weight did she lose from May to July?

   0 **A** 24 pounds

   0 **B** 14 pounds

   0 **C** 16 pounds

   0 **D** 170 pounds

2. Ross picked 56 cups of berries. Joe picked 34 cups. How many cups of berries did they pick?

   0 **A** 92 cups

   0 **B** 22 cups

   0 **C** 90 cups

   0 **D** 80 cups

3. Ramón has two fish tanks. One holds 52 gallons of water. The other holds 60 gallons. How much water does Ramón need to fill both tanks?

   0 **A** 102 gallons

   0 **B** 118 gallons

   0 **C** 138 gallons

   0 **D** 112 gallons

4. On his bus route, Mr. Kenny drives 23 miles in the morning and 32 miles in the afternoon. How many more miles does he drive in the afternoon than morning?

   0 **A** 55

   0 **B** 11

   0 **C** 9

   0 **D** 19

5. Linda used 82 inches of lace on a dress and 65 inches of lace on a skirt. How much more lace did she use on the dress than the skirt?

   0 **A** 17 inches

   0 **B** 147 inches

   0 **C** 27 inches

   0 **D** 23 inches

6. A club had 534 members in May and 618 members in June. How many more members did the club have in June?

   0 **A** 114

   0 **B** 74

   0 **C** 184

   0 **D** 84

Practice 6.A3

VI.A Select and use addition or subtraction to solve problems

1. Lynn had math and reading homework. She worked for 55 minutes. She spent 28 minutes on reading. How many minutes did she spend on math?

   0 **A** 27 minutes

   0 **B** 83 minutes

   0 **C** 33 minutes

   0 **D** 37 minutes

2. Mrs. Hummel drove 46 miles to visit her mother. She drove 28 more miles to see a friend. How many miles did she drive?

   0 **A** 18 miles

   0 **B** 22 miles

   0 **C** 64 miles

   0 **D** 74 miles

3. A park has 33 oak trees and 46 elm trees. How many oak and elm trees are in the park?

   0 **A** 71

   0 **B** 13

   0 **C** 79

   0 **D** 89

4. Mr. George bought 28 gallons of gasoline last week. This week he bought 34 gallons. How many more gallons did he buy this week than last week?

   0 **A** 14 gallons

   0 **B** 6 gallons

   0 **C** 62 gallons

   0 **D** 16 gallons

5. Misty made 80 cups of fruit punch for a party. At the end of the party, 13 cups of punch were left. How much punch did people drink at the party?

   0 **A** 77 cups

   0 **B** 93 cups

   0 **C** 67 cups

   0 **D** 77 cups

6. A football at the school store costs 225 points. Craig has 75 points. How many more points does he need for the football?

   0 **A** 255

   0 **B** 140

   0 **C** 300

   0 **D** 150

Practice 6.B1

VI.B Select and use multiplication (one-digit multiplier) or division (one-digit divisor) to solve problems

1. Mrs. Wing made 24 cookies for 6 students in her class. She gave the same number of cookies to each student. How many cookies did she give to each student?

   0 **A** 3

   0 **B** 4

   0 **C** 8

   0 **D** 144

2. Each student in Mr. Dotson's class earned 3 medals on field day. There are 15 students in the class. How many medals did they win in all?

   0 **A** 5

   0 **B** 6

   0 **C** 35

   0 **D** 45

3. Irene has 18 sheets of art paper. She uses 3 sheets to make one party hat. How many party hats can she make?

   0 **A** 54

   0 **B** 7

   0 **C** 6

   0 **D** 44

4. One cake serves 4 people. How many cakes will Becky need to serve 28 people?

   0 **A** 100

   0 **B** 8

   0 **C** 112

   0 **D** 7

5. A room has 14 rows of chairs. Each row has 7 chairs. How many chairs are in the room?

   0 **A** 98

   0 **B** 88

   0 **C** 2

   0 **D** 96

Practice 6.B2

VI.B Select and use multiplication (one-digit multiplier) or division (one-digit divisor) to solve problems

1. Sally made 32 brownies and stored them in 4 cans. She put the same number of brownies in each can. How many brownies were in each can?

   0 **A** 118

   0 **B** 9

   0 **C** 8

   0 **D** 128

2. Mr. Eliot had 10 flower pots. He planted 5 flower seeds in each pot. How many flower seeds did Mr. Eliot use?

   0 **A** 50

   0 **B** 15

   0 **C** 2

   0 **D** 40

3. Jaren usually scores 12 points in each basketball game. Last month he played in 4 games. How many points did he probably score?

   0 **A** 56

   0 **B** 3

   0 **C** 60

   0 **D** 48

4. Mrs. Robinson can drive 21 miles on one gallon of gas. How far can she drive on 3 gallons of gas?

   0 **A** 93 miles

   0 **B** 24 miles

   0 **C** 63 miles

   0 **D** 7 miles

5. Mario has 30 baseball cards in an album. If each page of the album has 5 cards, how many pages are used?

   0 **A** 150

   0 **B** 6

   0 **C** 4

   0 **D** 160

Practice 6.B3

1. Avery had 12 cans of paint. Each can held 6 cups of paint. How many cups of paint did Avery have?

   0 **A** 2 cups

   0 **B** 78 cups

   0 **C** 62 cups

   0 **D** 72 cups

2. Sheila used 9 inches of ribbon to make one bow. If she has 27 inches of ribbon, how many bows can she make?

   0 **A** 243

   0 **B** 3

   0 **C** 10

   0 **D** 5

3. Antonio bought 36 packs of gum. Each pack had 3 pieces of gum. How many pieces of gum did Antonio have?

   0 **A** 12

   0 **B** 13

   0 **C** 108

   0 **D** 98

4. Mrs. Davis made 48 donuts and packed them in boxes. She put 4 donuts in each box. How many boxes did she use?

   0 **A** 162

   0 **B** 13

   0 **C** 12

   0 **D** 192

5. Mr. Duncan had 16 salt shakers. He poured 4 ounces of salt in each salt shaker. How much salt did he use?

   0 **A** 64 ounces

   0 **B** 6 ounces

   0 **C** 4 ounces

   0 **D** 48 ounces

Practice 6.C1

VI.C Measure to solve problems involving length, area, temperature, and time

1. What is the **perimeter** (distance around) the rectangle?

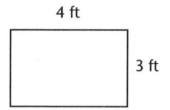

4 ft

3 ft

- 0 **A** 7 ft
- 0 **B** 8 ft
- 0 **C** 14 ft
- 0 **D** 16 ft

2. Look at the **shaded** part of the rectangle.

5 in

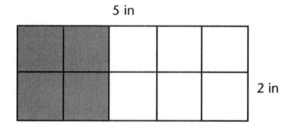

2 in

The **shaded** area is—

- 0 **A** 10 square inches
- 0 **B** 7 square inches
- 0 **C** 6 square inches
- 0 **D** 4 square inches

3. What temperature is shown on the thermometer?

- 0 **A** 72°F
- 0 **B** 70°F
- 0 **C** 68°F
- 0 **D** 66°F

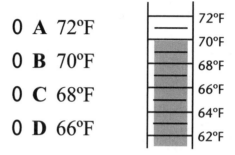

4. What time is shown on this clock?

- 0 **A** 9:15
- 0 **B** 9:25
- 0 **C** 9:30
- 0 **D** 9:45

5. What time will it be 15 minutes after the time shown on the clock?

- 0 **A** 5:50
- 0 **B** 6:15
- 0 **C** 6:20
- 0 **D** 6:30

Practice 6.C2

VI.C Measure to solve problems involving length, area, temperature, and time

1. What is the **perimeter** (distance around) this shape?

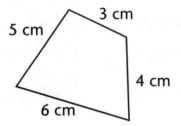

    0 **A** 12 cm

    0 **B** 13 cm

    0 **C** 15 cm

    0 **D** 18 cm

2. What is the **area** of the square?

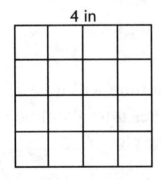

    0 **A** 20 square inches

    0 **B** 16 square inches

    0 **C** 12 square inches

    0 **D** 8 square inches

3. What temperature is shown on the thermometer?

    0 **A** 15°C

    0 **B** 17°C

    0 **C** 18°C

    0 **D** 19°C

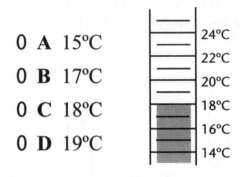

4. What time is shown on this clock?

    0 **A** 2:40

    0 **B** 8:00

    0 **C** 8:10

    0 **D** 8:20

5. Look at the clock. What time was it 15 minutes ago?

    0 **A** 4:00

    0 **B** 3:30

    0 **C** 3:15

    0 **D** 3:00

Practice 6.C3

VI.C Measure to solve problems involving length, area, temperature, and time

1. What is the **perimeter** (distance around) the triangle?

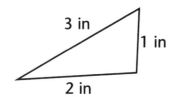

- 0 **A** 4 in
- 0 **B** 5 in
- 0 **C** 6 in
- 0 **D** 12 in

2. Look at the triangle inside the square.

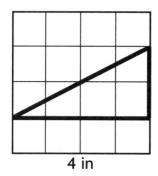

The **area** of the triangle is about—

- 0 **A** 2 square inches
- 0 **B** 4 square inches
- 0 **C** 12 square inches
- 0 **D** 16 square inches

3. What temperature is shown on the thermometer?

- 0 **A** 60°F
- 0 **B** 61°F
- 0 **C** 62°F
- 0 **D** 63°F

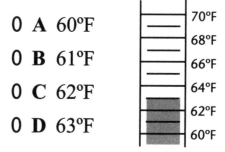

4. What time is shown on this clock?

- 0 **A** 5:55
- 0 **B** 6:00
- 0 **C** 6:05
- 0 **D** 6:15

5. What time will it be 25 minutes after the time shown on the clock?

- 0 **A** 4:35
- 0 **B** 5:05
- 0 **C** 5:15
- 0 **D** 5:25

Notes

Using Mathematical Representation

VII. **Express or solve problems using mathematical representation**

    A. Use appropriate mathematical notation and terms to express a solution

    B. Interpret information from pictographs and bar graphs

Notes

Objective 7: Pretest

1. A magazine costs 95¢. Jenny has 60¢. Her sister has 20¢. Which one shows how much money Jenny and her sister have together?

   0 **A** 60¢ + 20¢

   0 **B** 95¢ + 20¢

   0 **C** 95¢ + 60¢

   0 **D** 95¢ – 60¢

2. The Johnson family drove 110 miles on Monday morning. They drove 90 miles on Monday afternoon. On Tuesday they drove 300 miles. Which one shows how many miles they drove on Monday?

   0 **A** 300 – 90

   0 **B** 300 – 110

   0 **C** 110 + 300

   0 **D** 110 + 90

3. Hannah bought 13 gumballs and 10 lollipops. She gave 6 gumballs away. Which one shows how many gumballs she had then?

   0 **A** 13 + 10

   0 **B** 13 – 10

   0 **C** 13 – 6

   0 **D** 10 + 6

4. Cody found 15 leaves for his science project. His friend gave him 3 more. Which number sentence could you use to find how many leaves Cody had then?

   0 **A** $15 \div \square = 3$

   0 **B** $\square \times 3 = 15$

   0 **C** $15 + 3 = \square$

   0 **D** $15 - \square = 3$

5. Rick earned 8 tokens in math class. He wants a prize that costs 24 tokens. Which number sentence could you use to find how many more tokens he needs?

○ **A** $24 \div \square = 8$

○ **B** $24 - 8 = 16$

○ **C** $24 + 8 = 32$

○ **D** $8 \times \square = 24$

6. Jason bought 24 nails to make 3 tables. He used the same number of nails in each table. Which number sentence could you use to find how many nails he used in each table?

○ **A** $24 \times 3 = \square$

○ **B** $24 - \square = 3$

○ **C** $24 \div 3 = \square$

○ **D** $24 + 3 = \square$

7. Mrs. Clark has 20 students in her class. Each student checks out 2 books from the library. Which number sentence shows how to find the number of books the students checked out?

○ **A** $20 \times 2 = 40$

○ **B** $20 + 2 = 22$

○ **C** $20 \div 2 = 10$

○ **D** $20 - 2 = 18$

8. Juan had 10 baseball cards. He put the same number of cards in 2 boxes. Which number sentence shows how many cards he put in each box?

○ **A** $10 + 2 = 12$

○ **B** $10 - 2 = 8$

○ **C** $10 \times 2 = 20$

○ **D** $10 \div 2 = 5$

VII.B Interpret information from pictographs and bar graphs (9-16)

The graph shows the number of students who went to tutoring during one week. Use the graph to answer questions 9–12.

Students at Tutoring

Monday	✍ ✍ ✍ ✍ ✍
Tuesday	✍ ✍ ✍
Wednesday	✍ ✍ ✍ ✍
Thursday	✍ ✍ ✍ ✍ ✍ ✍
Friday	✍ ✍

Each ✍ = 1 student.

9. How many more students went to tutoring on Thursday than Tuesday?

   0 **A** 6

   0 **B** 5

   0 **C** 4

   0 **D** 3

10. Which is one way to find how many students went to tutoring on Monday, Wednesday, and Friday?

    0 **A** 5 + 3 + 4

    0 **B** 5 + 4 + 2

    0 **C** 9 + 3 + 2

    0 **D** 4 + 6 + 2

11. How many students went to tutoring on Wednesday and Thursday?

    0 **A** 4

    0 **B** 6

    0 **C** 8

    0 **D** 10

12. Which is one way to find how many more students went to tutoring on Thursday than on Friday?

    0 **A** 6 + 2

    0 **B** 6 − 4

    0 **C** 6 − 2

    0 **D** 6 + 3

The graph shows how much rain fell in Austin for five months. Use the graph to answer questions 13–16.

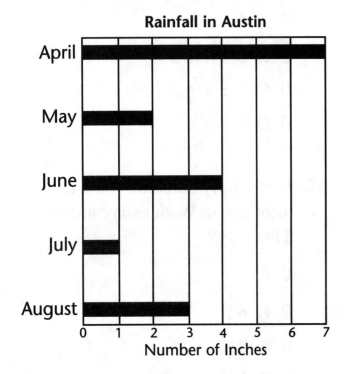

13. In which month did the least amount of rain fall?

  0 **A** August

  0 **B** July

  0 **C** June

  0 **D** May

14. How many inches of rain fell in June?

  0 **A** 1 inch

  0 **B** 2 inches

  0 **C** 4 inches

  0 **D** 7 inches

15. How many more inches of rain fell in April than in July?

  0 **A** 10 inches

  0 **B** 8 inches

  0 **C** 7 inches

  0 **D** 6 inches

16. Which is one way to find how much rain fell in June, July, and August?

  0 **A** 4 + 1 + 3

  0 **B** 3 + 4 + 3

  0 **C** 7 + 4 + 3

  0 **D** 7 + 2 + 4

Practice 7.A1

VII.A Use appropriate mathematical notation and terms to express a solution

1. At the school fair, 10 students rode the merry-go-round, and 8 students rode the Ferris wheel. Then 6 more students got on the Ferris wheel. Which shows how many students were on the Ferris wheel?

   0 **A** 10 + 8

   0 **B** 10 – 8

   0 **C** 10 + 6

   0 **D** 8 + 6

2. A bike race covers 50 miles. On the first day, Sunil rode 22 miles on his bike. Tony rode 15 miles on his bike. Which one shows how many more miles Sunil rode than Tony?

   0 **A** 50 + 22

   0 **B** 22 + 15

   0 **C** 50 – 22

   0 **D** 22 – 15

3. Mrs. Winn bought 6 boxes of pens. Each box had 12 pens. Which number sentence shows how many pens Mrs. Winn had?

   0 **A** 12 ÷ 6 = 2

   0 **B** 12 x 6 = 72

   0 **C** 12 – 6 = 6

   0 **D** 12 + 6 = 18

4. Rashid had 35¢. He found 7¢ on the playground. Which number sentence could be used to find how much Rashid had then?

   0 **A** 35¢ ÷ 7¢ = ☐

   0 **B** 35¢ – 7¢ = ☐

   0 **C** 35¢ + 7¢ = ☐

   0 **D** 35¢ x 7¢ = ☐

5. Dale has 25 more marbles than Andy. Andy has 5 marbles. Which number sentence could be used to find how many marbles Dale has?

   0 **A** 25 + 5 = ☐

   0 **B** 25 – 5 = ☐

   0 **C** 25 ÷ 5 = ☐

   0 **D** 25 x 5 = ☐

Practice 7.A2

VII.A Use appropriate mathematical notation and terms to express a solution

1. Jan has 8 more stamps than Larry. Larry has 64 stamps. Which number sentence shows how many stamps Jan has?

   0 **A** $64 - 8 = 56$

   0 **B** $64 \div 8 = 8$

   0 **C** $64 + 8 = 72$

   0 **D** $64 \times 8 = 512$

2. Joe had 10 pieces of gum. He gave an equal amount of gum to 5 friends. Which number sentence could be used to find how many pieces of gum he gave to each student?

   0 **A** $10 \div 5 = \square$

   0 **B** $10 \times 5 = \square$

   0 **C** $10 - 5 = \square$

   0 **D** $10 + 5 = \square$

3. Edna asked 4 friends to her party. She gave each friend 4 party favors. Which one shows how many party favors Edna gave?

   0 **A** $4 + 4$

   0 **B** 4×4

   0 **C** $4 - 4$

   0 **D** $4 \div 4$

4. There were 27 visitors at a museum. They took tours in groups of 3. Which number sentence could be used to find how many groups they made?

   0 **A** $27 \times 3 = \square$

   0 **B** $27 - 3 = \square$

   0 **C** $27 + 3 = \square$

   0 **D** $27 \div 3 = \square$

5. Jill drinks 16 cups of juice each week. Which number sentence could be used to find how many cups she drinks in 4 weeks?

   0 **A** $16 \times 4 = \square$

   0 **B** $16 - 4 = \square$

   0 **C** $16 \div 4 = \square$

   0 **D** $16 + 4 = \square$

Practice 7.A3

VII.A *Use appropriate mathematical notation and terms to express a solution*

1. Angie walks 4 miles every day. Which one shows how many miles she walks in 7 days?

   0 **A** $4 + 7$

   0 **B** 4×7

   0 **C** $7 - 4$

   0 **D** $7 \div 4$

2. Randy earned 3 points for each book he read. He earned 18 points. Which number sentence shows how many books he read?

   0 **A** $18 - 3 = 15$

   0 **B** $18 \times 3 = 54$

   0 **C** $18 + 3 = 21$

   0 **D** $18 \div 3 = 6$

3. Mr. Malone drove 300 miles on a trip. The first day he drove 100 miles. Which number sentence could be used to find how many more miles he had to drive?

   0 **A** $300 \times 100 = \square$

   0 **B** $300 - 100 = \square$

   0 **C** $300 + 100 = \square$

   0 **D** $300 \div 100 = \square$

4. Bonnie gave 3 stickers to each student in her class. She gave stickers to 30 students. Which one could be used to find how many stickers she gave?

   0 **A** $30 \times 3 = \square$

   0 **B** $30 - 3 = \square$

   0 **C** $30 + 3 = \square$

   0 **D** $30 \div 3 = \square$

5. Bob sold 20 candy bars to his aunt and 5 candy bars to his uncle. Which number sentence shows how many candy bars he sold?

   0 **A** $20 - 5 = 15$

   0 **B** $20 \times 5 = 100$

   0 **C** $20 + 5 = 25$

   0 **D** $20 \div 5 = 4$

Practice 7.B1

VII.B Interpret information from pictographs and bar graphs

The graph shows the number of pets owned by five families. Use the graph to answer questions 1–4.

Pets Owned

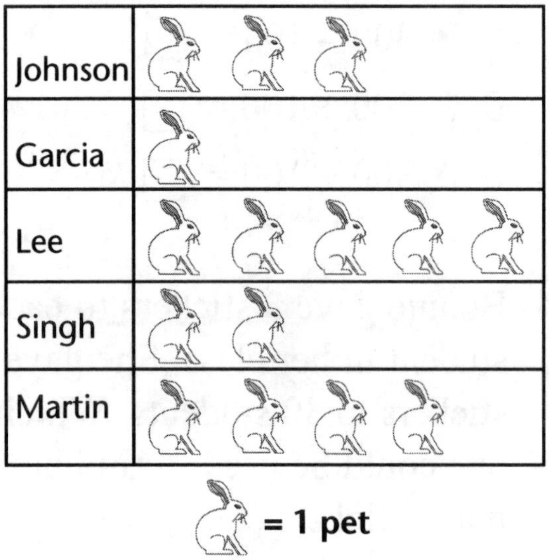

= 1 pet

1. How many more pets does the Lee family own than the Singh family?

    0 **A** 7

    0 **B** 5

    0 **C** 4

    0 **D** 3

2. Which is one way to find the total number of pets owned by the Johnson, Garcia, and Lee families?

    0 **A** 3 + 1 + 5

    0 **B** 3 + 6 + 2

    0 **C** 5 + 2 + 4

    0 **D** 9 + 2 + 4

3. Which family owns 4 pets?

    0 **A** Johnson

    0 **B** Lee

    0 **C** Singh

    0 **D** Martin

4. Which is one way to find how many more pets the Martin family has than the Johnson family?

    0 **A** 4 + 3

    0 **B** 5 − 1

    0 **C** 4 − 3

    0 **D** 5 + 3

Practice 7.B2

VII.B Interpret information from pictographs and bar graphs

The graph shows the number of students in Mrs. King's class who play different sports. Use the graph to answer questions 1–4.

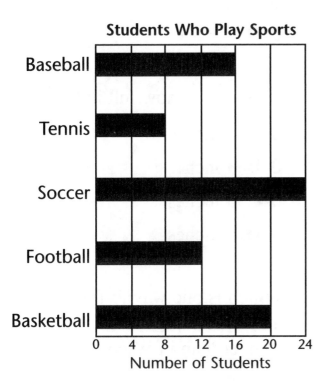

Students Who Play Sports

1. The greatest number of students in Mrs. King's class play—

   0 **A** baseball

   0 **B** soccer

   0 **C** football

   0 **D** tennis

2. How many more students play basketball than tennis?

   0 **A** 3

   0 **B** 6

   0 **C** 12

   0 **D** 28

3. Which is one way to find the number of students that play football or baseball?

   0 **A** 16 + 8

   0 **B** 16 – 12

   0 **C** 8 + 6

   0 **D** 16 + 12

4. Which is one way to find how many more students play soccer than tennis?

   0 **A** 24 + 8

   0 **B** 24 – 16

   0 **C** 24 – 8

   0 **D** 12 – 4

Practice 7.B3

VII.B *Interpret information from pictographs and bar graphs*

The graph shows how much snow fell in five cities. Use the graph to answer questions 1–4.

Snowfall in Five Cities

Dayton	❄ ❄
Denver	❄ ❄ ❄ ❄ ❄
Gary	❄ ❄ ❄
New York	❄ ❄ ❄ ❄
Waco	❄

Each ❄ = 2 inches.

1. How much more snow fell in New York than in Waco?

    0 **A** 3 inches

    0 **B** 5 inches

    0 **C** 6 inches

    0 **D** 10 inches

2. Which is one way to find the total amount of snowfall in Dayton and Denver?

    0 **A** 2 + 5

    0 **B** 4 + 10

    0 **C** 2 + 10

    0 **D** 4 + 5

3. How much snow fell in Gary?

    0 **A** 2 inches

    0 **B** 3 inches

    0 **C** 5 inches

    0 **D** 6 inches

4. Which is one way to find how much more snow fell in Denver than in Waco?

    0 **A** 5 + 1

    0 **B** 10 − 2

    0 **C** 10 + 2

    0 **D** 10 − 1

Practice 7.B4

VII.B   Interpret information from pictographs and bar graphs

The graph shows how many students went to tutoring during one week. Use the graph to answer questions 1–4.

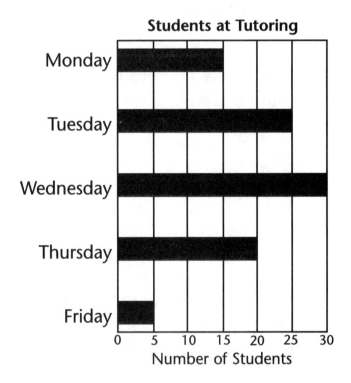

Students at Tutoring

Number of Students

1. How many more students went to tutoring on Wednesday than on Thursday?

   0 **A** 30

   0 **B** 20

   0 **C** 15

   0 **D** 10

2. Which is one way to find the number of students who went to tutoring on Monday and Tuesday?

   0 **A** 25 – 15

   0 **B** 25 + 15

   0 **C** 25 + 30

   0 **D** 20 – 5

3. How many students went to tutoring on Thursday?

   0 **A** 10

   0 **B** 15

   0 **C** 20

   0 **D** 25

4. Which is one way to find how many more students went to tutoring on Monday than on Friday?

   0 **A** 15 + 5

   0 **B** 15 – 5

   0 **C** 25 – 5

   0 **D** 20 + 5

107

Practice 7.B5

VII.B Interpret information from pictographs and bar graphs

The graph shows how many families live on five streets in Newton City. Use the graph to answer questions 1–4.

Number of Families Living on Five Streets

Main St.	◆ ◆ ◆ ◆ ◆
Elm St.	◆ ◆ ◆ ◆ ◆ ◆
High St.	◆ ◆ ◆ ◆
Oak St.	◆ ◆ ◆ ◆ ◆ ◆
Clark St.	◆ ◆ ◆

Each ◆ = 5 families.

1. How many more families live on High Street than on Clark Street?

0 **A** 5

0 **B** 10

0 **C** 15

0 **D** 35

2. On which two streets do the same number of families live?

0 **A** Main and High

0 **B** Main and Clark

0 **C** Elm and Oak

0 **D** Elm and High

3. How many families live on Main Street?

0 **A** 5

0 **B** 10

0 **C** 20

0 **D** 25

4. Which is one way to find how many more families live on Oak Street than on Main Street?

0 **A** 25 + 30

0 **B** 30 – 25

0 **C** 6 – 5

0 **D** 6 x 5

Practice 7.B6

VII.B Interpret information from pictographs and bar graphs

The graph shows how much money five teams earned for a school fund-raiser. Use the graph to answer questions 1–4.

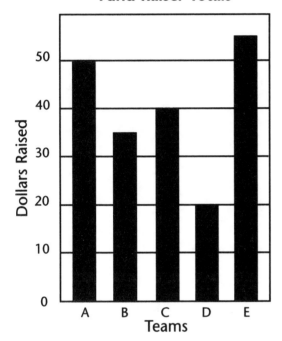

Fund-Raiser Totals

1. Which team earned about $35?

   0 **A** Team A

   0 **B** Team B

   0 **C** Team C

   0 **D** Team D

2. How much more did Team A make than Team C?

   0 **A** $10

   0 **B** $15

   0 **C** $20

   0 **D** $90

3. About how much money did Team E earn?

   0 **A** $40

   0 **B** $45

   0 **C** $50

   0 **D** $55

4. Which is one way to find how much Team B and Team D earned together?

   0 **A** $35 – $20

   0 **B** $35 + $40

   0 **C** $35 + $20

   0 **D** $20 + $40

Notes

©ECS Learning Systems, Inc.

110

Evaluating Reasonableness

VIII. Evaluate the reasonableness of a solution to a problem situation

    A. Use a problem-solving model that incorporates understanding the problem, making a plan, carrying out the plan, and evaluating the solution for reasonableness

Notes

Objective 8: Pretest

VIII.A Use problem-solving skills to solve problems and evaluate reasonableness of solution

1. Teddy was counting desks for the principal. What is a reasonable number of desks he might have counted in 5 average classrooms?

   0 **A** 500

   0 **B** 150

   0 **C** 30

   0 **D** 10

2. Randy and Charles are third-grade boys. About how much would two third-grade boys weigh together?

   0 **A** 12,000 pounds

   0 **B** 1,200 pounds

   0 **C** 120 pounds

   0 **D** 40 pounds

3. Marcy bought 6 gifts. The lowest-priced gift was $3. The highest-priced gift was $10. What is a reasonable total for the cost of all 6 gifts?

   0 **A** more than $60

   0 **B** between $32 and $40

   0 **C** between $10 and $20

   0 **D** less than $10

4. Sherry will serve cake to 18 guests. Each cake serves 5 people. How many cakes does Sherry need?

   0 **A** 6

   0 **B** 4

   0 **C** 3

   0 **D** 2

5. If you weighed one pair of your shoes, about how much would you expect them to weigh?

   0 **A** 2,000 pounds

   0 **B** 200 pounds

   0 **C** 20 pounds

   0 **D** 2 pounds

6. Lucy made hats for her 6 dolls. She needed 4 inches of lace for each hat. What is a reasonable amount of lace for Lucy to buy?

   0 **A** 10 inches

   0 **B** 15 inches

   0 **C** 25 inches

   0 **D** 50 inches

Practice 8.A1

1. Philip bought 7 watermelons. Each one cost $2. What is a reasonable amount of money that Philip might give to the sales clerk?

   0 **A** $100

   0 **B** $50

   0 **C** $30

   0 **D** $15

2. If you counted the number of pages in your science book, about how many pages might there be?

   0 **A** 14

   0 **B** 40

   0 **C** 240

   0 **D** 4,000

3. Crissy bought 4 things at the store. The lowest-priced item was $1. The highest-priced item was $5. What is a reasonable total that Crissy paid for all the items?

   0 **A** less than $5

   0 **B** between $5 and $7

   0 **C** between $8 and $16

   0 **D** more than $20

4. Mrs. Fine wants to make 20 sandwiches. One package of meat will make 8 sandwiches. How many packages of meat should Mrs. Fine buy?

   0 **A** 1

   0 **B** 3

   0 **C** 8

   0 **D** 16

5. Ed jogged on 4 days. He ran at least 5 miles every day. One day he ran 7 miles. What is a reasonable total for the number of miles Ed jogged?

   0 **A** less than 8 miles

   0 **B** between 8 and 18 miles

   0 **C** between 22 and 28 miles

   0 **D** more than 30 miles

Practice 8.A2

1. Ellen bought 3 apples at the grocery store. Which one is a reasonable weight for 3 apples?

   0 **A** 20 pounds

   0 **B** 10 pounds

   0 **C** 5 pounds

   0 **D** 1 pound

2. Heidi walks from 4 to 6 miles each week. What is a reasonable number of miles she might walk in 6 weeks?

   0 **A** more than 50

   0 **B** between 40 and 50 miles

   0 **C** between 24 and 36 miles

   0 **D** less than 20 miles

3. Tony uses 1 cup of sugar in each batch of cookies. One bag of sugar holds 2 cups. If Tony makes 5 batches of cookies, how many bags of sugar should he buy?

   0 **A** 1

   0 **B** 2

   0 **C** 3

   0 **D** 5

4. Selena measured the **perimeter** (distance around) her bedroom. Which is a reasonable estimate of that distance?

   0 **A** 4 feet

   0 **B** 40 feet

   0 **C** 400 feet

   0 **D** 4,000 feet

5. Shirts cost between $15 and $20 at a store. Joe buys 4 shirts. What is a reasonable amount he will pay for the shirts?

   0 **A** less than $40

   0 **B** between $40 and $50

   0 **C** between $60 and $80

   0 **D** more than $100

6. Mr. Ryan has 60 pounds of apples to put in baskets. One basket can hold no more than 8 pounds of apples. How many baskets will Mr. Ryan have to use for all the apples?

   0 **A** 5

   0 **B** 6

   0 **C** 7

   0 **D** 8

Practice 8.A3

1. Nate earns from $2 to $4 each week by doing chores. Which is a reasonable amount that he will have made at the end of 4 weeks?

   0 **A** less than $5

   0 **B** between $8 and $16

   0 **C** between $16 and $20

   0 **D** more than $20

2. A pack of gum holds 5 pieces. Janie chews 2 pieces of gum each day. About how many packs of gum would she use in a week (7 days)?

   0 **A** 7

   0 **B** 5

   0 **C** 3

   0 **D** 2

3. Kathy bought 1 pound of oranges. About how many oranges did she probably get?

   0 **A** 100

   0 **B** 50

   0 **C** 15

   0 **D** 5

4. It takes Jan about 20 minutes to walk from her house to school. What is a reasonable estimate of the distance she walks?

   0 **A** 1 mile

   0 **B** 5 miles

   0 **C** 10 miles

   0 **D** 20 miles

5. One package of chips serves 6 people. You serve chips to 15 people. What is a reasonable number of packages you used?

   0 **A** 6

   0 **B** 5

   0 **C** 3

   0 **D** 1

6. Carrie made 13 book covers. For each cover she used a little less than 3 feet of paper. Which is a reasonable estimate of the amount of paper she used?

   0 **A** 20 feet

   0 **B** 40 feet

   0 **C** 80 feet

   0 **D** 100 feet

Practice 8.A4

VIII.A Use problem-solving skills to solve problems and evaluate reasonableness of solution

1. Marty watches TV from 2 to 3 hours every day. What is a reasonable number of hours he might watch in 10 days?

   O **A** more than 50 hours

   O **B** between 35 and 50 hours

   O **C** between 20 and 30 hours

   O **D** less than 20 hours

2. Mrs. Martini wants to make 44 cupcakes. One package of cake mix will make 12 cupcakes. How many packages of cake mix will Mrs. Martini use?

   O **A** 3

   O **B** 4

   O **C** 6

   O **D** 8

3. Movie tickets cost from $5 to $7. What is a reasonable amount you might pay to see 5 different movies?

   O **A** less than $20

   O **B** between $25 and $35

   O **C** between $40 and $50

   O **D** more than $50

4. It takes Jenny about 7 minutes to wrap one present. What is a reasonable amount of time she would spend to wrap 8 presents?

   O **A** 20 minutes

   O **B** 40 minutes

   O **C** 60 minutes

   O **D** 100 minutes

5. A waiter can pour 6 glasses of tea from one pitcher. How many times would he have to refill the pitcher to serve 20 glasses of tea?

   O **A** 10

   O **B** 8

   O **C** 5

   O **D** 4

6. Doug can read less than 2 pages of a story in one minute. What is a reasonable amount of time it would take him to read a 13-page story?

   O **A** 60 minutes

   O **B** 30 minutes

   O **C** 10 minutes

   O **D** 5 minutes

Notes

Appendix

- **Answer Key**
- **Answer Sheets**

Notes

Answer Key: Using
Addition to Solve Problems

Objective 1 Pretest (p. 11)

1. D	2. C	3. D	4. B	5. A
6. C	7. D	8. C	9. B	10. C
11. C	12. B	13. C	14. D	15. C
16. A	17. C	18. C	19. D	20. D

Practice 1.A1 (p. 15)

1. D	2. B	3. C	4. B	5. D

Practice 1.A2 (p. 16)

1. C	2. A	3. A	4. C	5. D

Practice 1.A3 (p. 17)

1. C	2. D	3. D	4. B	5. D

Practice 1.A4 (p. 18)

1. D	2. C	3. A	4. B	5. C
6. B				

Practice 1.B1 (p. 19)

1. D	2. D	3. A	4. A	5. C
6. C				

Practice 1.B2 (p. 20)

1. D	2. A	3. C	4. C	5. D
6. C				

Practice 1.B3 (p. 21)

1. B	2. C	3. D	4. D	5. C
6. B				

Practice 1.B4 (p. 22)

1. D	2. B	3. D	4. D	5. A
6. B				

Practice 1.C1 (p. 23)

1. B	2. C	3. C	4. D	5. B

Practice 1.C2 (p. 24)

1. C	2. B	3. D	4. C	5. B

Practice 1.C3 (p. 25)

1. D	2. C	3. A	4. C	5. C

Practice 1.C4 (p. 26)

1. B	2. C	3. C	4. A	5. D

Practice 1.D1 (p. 27)

1. D	2. B	3. C	4. B

Practice 1.D2 (p. 28)

1. D	2. B	3. D	4. B

Practice 1.D3 (p. 29)

1. D	2. B	3. C	4. C	5. B

Practice 1.D4 (p. 30)

1. B	2. D	3. A	4. A	5. C

Answer Key: Using
Subtraction to Solve Problems

Objective 2 Pretest (p. 33)

1. C	2. A	3. C	4. D	5. D
6. B	7. C	8. D	9. C	10. D
11. A	12. D	13. C	14. B	15. D

Practice 2.A1 (p. 36)

1. D	2. A	3. C	4. D	5. D

Practice 2.A2 (p. 37)

1. D	2. B	3. D	4. B	5. C

Practice 2.A3 (p. 38)

1. D	2. D	3. A	4. A	5. C

Practice 2.A4 (p. 39)

1. A	2. A	3. C	4. C	5. D

Practice 2.B1 (p. 40)

1. B	2. C	3. D	4. B	5. D
6. A				

121

Practice 2.B2 (p. 41)
1. C   2. D   3. A   4. C   5. B
6. D

Practice 2.B3 (p.42)
1. B   2. B   3. C   4. A   5. C
6. C

Practice 2.B4 (p. 43)
1. B   2. A   3. D   4. D   5. C
6. B

Practice 2.C1 (p. 44)
1. B   2. D   3. A   4. C   5. A

Practice 2.C2 (p. 45)
1. B   2. C   3. B   4. D   5. A

Practice 2.C3 (p. 46)
1. C   2. D   3. D   4. A   5. B

Answer Key: Using Multiplication to Solve Problems

Objective 3 Pretest (p. 49)
1. D   2. C   3. A   4. C   5. D
6. D   7. B   8. C   9. D   10. D
11. B   12. C   13. D   14. A   15. D
16. A   17. C   18. D

Practice 3.A1 (p. 52)
1. C   2. B   3. C   4. C   5. D
6. B

Practice 3.A2 (p. 53)
1. D   2. A   3. C   4. C   5. A
6. C

Practice 3.A3 (p. 54)
1. B   2. C   3. C   4. D   5. C
6. D

Practice 3.A4 (p. 55)
1. D   2. A   3. C   4. C   5. D
6. B

Answer Key: Using Division to Solve Problems

Objective 4 Pretest (p. 59)
1. C   2. B   3. D   4. C   5. B
6. D   7. A   8. D   9. A   10. D
11. C   12. A   13. C   14. B   15. C

Practice 4.A1 (p. 62)
1. D   2. C   3. A   4. B

Practice 4.A2 (p. 63)
1. C   2. D   3. B   4. D

Practice 4.A3 (p. 64)
1. B   2. D   3. D   4. C

Practice 4.B1 (p. 65)
1. D   2. B   3. D   4. C   5. D
6. A

Practice 4.B2 (p. 66)
1. D   2. B   3. A   4. A   5. B
6. C

Practice 4.B3 (p. 67)
1. A   2. A   3. C   4. C   5. B
6. B

Practice 4.C1 (p. 68)
1. C   2. D   3. B   4. B   5. A
6. C

Practice 4. C2 (p. 69)
1. B   2. C   3. B   4. A   5. C
6. C

Practice 4.C3 (p. 70)
1. D   2. C   3. B   4. C   5. D
6. B

Answer Key: Estimate Solutions

Objective 5 Pretest (p. 73)
1. D   2. B   3. C   4. B   5. A
6. D   7. C   8. A   9. B   10. D
11. D   12. A

Practice 5.A1 (p. 75)
1. C   2. C   3. D   4. B   5. D
6. A

Practice 5.A2 (p. 76)
1. C   2. B   3. A   4. C   5. A
6. D

Practice 5.A3 (p. 77)
1. D   2. A   3. C   4. C   5. B
6. D

Practice 5.A4 (p. 78)
1. C   2. B   3. C   4. A   5. D
6. B

Answer Key: Solution Strategies/Analyze or Solve Problems

Objective 6 Pretest (p. 81)
1. D   2. A   3. D   4. C   5. B
6. D   7. A   8. D   9. B   10. C
11. C   12. C   13. D   14. A   15. B
16. A   17. B   18. D

Practice 6.A1 (p. 85)
1. D   2. C   3. A   4. D   5. A
6. B

Practice 6.A2 (p. 86)
1. B   2. C   3. D   4. C   5. A
6. D

Practice 6.A3 (p. 87)
1. A   2. D   3. C   4. B   5. C
6. D

Practice 6.B1 (p. 88)
1. B   2. D   3. C   4. D   5. A

Practice 6.B2 (p. 89)
1. C   2. A   3. D   4. C   5. B

Practice 6.B3 (p. 90)
1. D   2. B   3. C   4. C   5. A

Practice 6.C1 (p. 91)
1. C   2. D   3. B   4. B   5. C

Practice 6.C2 (p. 92)
1. D   2. B   3. C   4. C   5. B

Practice 6.C3 (p. 93)
1. C   2. B   3. D   4. C   5. D

Answer Key: Using Mathematical Representation

Objective 7 Pretest (p. 97)
1. A   2. D   3. C   4. C   5. B
6. C   7. A   8. D   9. D   10. B
11. D   12. C   13. B   14. C   15. D
16. A

Practice 7.A1 (p. 101)
1. D   2. D   3. B   4. C   5. A

Practice 7.A2 (p. 102)
1. C   2. A   3. B   4. D   5. A

Practice 7.A3 (p. 103)
1. B 2. D 3. B 4. A 5. C

Practice 7.B1 (p. 104)
1. D 2. A 3. D 4. C

Practice 7.B2 (p. 105)
1. B 2. C 3. D 4. C

Practice 7.B3 (p. 106)
1. C 2. B 3. D 4. B

Practice 7.B4 (p. 107)
1. D 2. B 3. C 4. B

Practice 7.B5 (p. 108)
1. A 2. C 3. D 4. B

Practice 7.B6 (p. 109)
1. B 2. A 3. D 4. C

Answer Key: Evaluating Reasonableness

Objective 8 Pretest (p. 113)
1. B 2. C 3. B 4. B 5. D
6. C

Practice 8.A1 (p. 114)
1. D 2. C 3. C 4. B 5. C

Practice 8.A2 (p. 115)
1. D 2. C 3. C 4. B 5. C
6. D

Practice 8.A3 (p. 116)
1. B 2. C 3. D 4. A 5. C
6. B

Practice 8.A4 (p. 117)
1. C 2. B 3. B 4. C 5. D
6. C

Name _____   **Date** _____

Objective # _____ Pretest

Pretest Answer Sheet

1. Ⓐ Ⓑ Ⓒ Ⓓ   11. Ⓐ Ⓑ Ⓒ Ⓓ

2. Ⓐ Ⓑ Ⓒ Ⓓ   12. Ⓐ Ⓑ Ⓒ Ⓓ

3. Ⓐ Ⓑ Ⓒ Ⓓ   13. Ⓐ Ⓑ Ⓒ Ⓓ

4. Ⓐ Ⓑ Ⓒ Ⓓ   14. Ⓐ Ⓑ Ⓒ Ⓓ

5. Ⓐ Ⓑ Ⓒ Ⓓ   15. Ⓐ Ⓑ Ⓒ Ⓓ

6. Ⓐ Ⓑ Ⓒ Ⓓ   16. Ⓐ Ⓑ Ⓒ Ⓓ

7. Ⓐ Ⓑ Ⓒ Ⓓ   17. Ⓐ Ⓑ Ⓒ Ⓓ

8. Ⓐ Ⓑ Ⓒ Ⓓ   18. Ⓐ Ⓑ Ⓒ Ⓓ

9. Ⓐ Ⓑ Ⓒ Ⓓ   19. Ⓐ Ⓑ Ⓒ Ⓓ

10. Ⓐ Ⓑ Ⓒ Ⓓ   20. Ⓐ Ⓑ Ⓒ Ⓓ

Practice Answer Sheet

Name _____

Date: _____ **Practice: #** _____

1. Ⓐ Ⓑ Ⓒ Ⓓ

2. Ⓐ Ⓑ Ⓒ Ⓓ

3. Ⓐ Ⓑ Ⓒ Ⓓ

4. Ⓐ Ⓑ Ⓒ Ⓓ

5. Ⓐ Ⓑ Ⓒ Ⓓ

6. Ⓐ Ⓑ Ⓒ Ⓓ

Date: _____ **Practice: #** _____

1. Ⓐ Ⓑ Ⓒ Ⓓ

2. Ⓐ Ⓑ Ⓒ Ⓓ

3. Ⓐ Ⓑ Ⓒ Ⓓ

4. Ⓐ Ⓑ Ⓒ Ⓓ

5. Ⓐ Ⓑ Ⓒ Ⓓ

6. Ⓐ Ⓑ Ⓒ Ⓓ

Date: _____ **Practice: #** _____

1. Ⓐ Ⓑ Ⓒ Ⓓ

2. Ⓐ Ⓑ Ⓒ Ⓓ

3. Ⓐ Ⓑ Ⓒ Ⓓ

4. Ⓐ Ⓑ Ⓒ Ⓓ

5. Ⓐ Ⓑ Ⓒ Ⓓ

6. Ⓐ Ⓑ Ⓒ Ⓓ

Date: _____ **Practice: #** _____

1. Ⓐ Ⓑ Ⓒ Ⓓ

2. Ⓐ Ⓑ Ⓒ Ⓓ

3. Ⓐ Ⓑ Ⓒ Ⓓ

4. Ⓐ Ⓑ Ⓒ Ⓓ

5. Ⓐ Ⓑ Ⓒ Ⓓ

6. Ⓐ Ⓑ Ⓒ Ⓓ